Synthetic, Imitation and Treated Gemstones

Contents

Preface

Three pictures

Early in 1996 a jewellery catalogue from a major auction house featured jewellery set with blue Maxixe beryls. These stones, rivalling the best blue sapphires in depth and beauty of colour but not entirely sapphire-like, were first discovered in the early years of this century. Sadly the stones did not maintain their magnificent colour but slowly faded to a pale pink to straw yellow.

In the 1970s, similar blue beryls hit the London and other markets, and their colour faded similarly. The stones disappeared from the markets, and many London and German dealers assured me that 'they had not been taken in and did not buy the stones'. If this was true, where did I hear the quoted price of £85 a carat, which, as I have kept price records for years, I noted down at the time?

When I first began teaching gemmology (gem testing) at London Guildhall University in the late 1960s the textbooks of the day, with the confidence characteristic of what was then a largely amateur-driven study, stated that synthetic gem diamond 'would not appear in the diamond market in our time'. Perhaps not. But here it is today: in 1995 a yellow diamond sent to the Gemmological Association and Gem Testing Laboratory of Great Britain for grading was found to be *synthetic*.

While working in the sapphire fields of Montana, USA, in 1992 I found many crystals with a pale green, blue, yellow or orange colour. It was then known that many such specimens would respond to heating by producing a much stronger version of their original colour, and since that time I have examined many hundreds of Montana sapphire specimens. They display very fine, strong colours and are very suitable for both classical-style and modern jewellery, being routinely sold as heated stones in the USA.

A common thread links the three pictures. In all cases the stones concerned were not entirely what they seemed to be. The original Maxixe beryls were sold in ignorance of their propensities, but those sold much later were treated to obtain their fine blue colour. The diamond was laboratory (or perhaps even factory), made and the Montana sapphires had no commercial value until they had been heated to give a permanent and enhanced colour.

It is said that almost all emeralds placed on the market have been oiled to improve their colour: emeralds and rubies have had glassy and plastic materials inserted into cracks, and diamonds may have unsightly inclusions improved by lasering. We shall be looking at all these practices, but we must always remember that the customer, upon whom the jewellery trade relies, needs constant assurance that his or her purchases are not only entirely natural but capable of being sold on at a profit. It is here that the whole question of manufacture and treatment has become an immediate concern to everyone concerned with gemstones. Knowledge is the key to successful gemstone and jewellery selling, and customers are ever more demanding, as they have every right to be.

What was until recently a largely academic study for gemmologists now directly affects the jewellery-buying public, who need to be served by much better-educated sales staff.

I hope that this book will help in the education process.

The author is most grateful to the Gemmological Association and Gem Testing Laboratory of Great Britain for the loan of photographs. Thanks are also due to Sotheby's International SA and David Bennett for permission to use the photograph of blue Maxixe beryl's set in jewellery on the front cover of the book.

Michael O'Donoghue

Chapter 1

Introduction

Gemmologists, jewellers, gemstone and mineral collectors, museum curators and even gem and mineral prospectors now accept that nothing is what it seems. Perhaps it never was! Today a gemstone may be natural, natural but with improved colour, an entirely artificial substance with no natural counterpart, a man-made material with a natural counterpart, a composite stone partly natural and partly artificial or entirely one or the other. While testing an unknown to prove its identity is one problem, another is 'what to call it'. The question of nomenclature has been with us for the whole of this century, and seems no nearer to resolution as we approach the millennium. As the book proceeds we shall see that the problem is not easy since the meaning of words varies from language to language and even within English.

This book is not intended as a textbook in elementary gemmology and will not include details of crystal systems and of the theory behind the operation of standard gem testing instruments. None the less, some knowledge of instruments is necessary for anyone needing to prove the nature of a specimen and brief explanations will be provided when the context calls for them. On the other hand the book is not intended to be a history of the development of gem crystal growth, and it certainly is not a guide to heating, irradiating, oiling, infilling, lasering or even making your own gemstones!

Nomenclature

This nettle has to be grasped from the outset. The present practice is to call natural, untreated stones 'natural' and to use the adjective 'synthetic' for man-made substances which have a natural counterpart, even if these are mineral species of very limited size and occurrence. Confusion has often been caused by the adjective 'imitation': while it has traditionally been used to describe substances which imitate the major gem species, and in particular, glass, there is no doubt that any material can imitate another, whatever its origin. To a possible collector of colourless synthetic spinel who also wants specimens which resemble it, diamond itself will be an imitation! There is a difference in usage between the general public and jewellers, even though few jewellers would offer anything as an 'imitation' – far rather would they make up some grand-sounding name, since they have to sell their goods. Gemmologists in trying to remain true to what they see as the usage of descriptive science have accidentally limited the field in which 'imitation' can be used. Perhaps the term 'artificial' could take a useful place: thus allowing man-made substances to be grouped together and the 'natural counterpart' proviso to be abandoned. Since most jewellers and their customers would not be aware of the obscurer counterparts, this could be a useful new blanket term.

Chapter 2

Plan of the book

The book is arranged by species, beginning with diamond and passing from ruby, sapphire, emerald and opal to less well-known but equally beautiful species. Within the species we first discuss the basic nature of the material, just sufficiently to allow the jeweller or the gemmologist with a 10× lens or simple instruments which can be used with little or no previous knowledge of gem testing to appreciate the phenomena observed.

The 10× lens has its limitations, however and the second part of the survey of each gem species looks at the ways they can be tested by the standard gemmological instruments and the routine ways in which the instruments are used, though no attempt is made to explain any but the basic principles behind their operation. A separate chapter covers instruments in greater detail for those who want to know more about them.

Some topics, especially those involving the establishment of the colour of a gemstone as natural or artificially produced, need more sophisticated testing: this can be provided only by dedicated gemmological laboratories or by facilities in major universities or museums. Such tests are described but in general only their results will be given.

In looking at the major gem species, we first need to establish the identity of those natural and artificial materials that imitate them and their nature. We also have to examine synthetic counterparts of natural stones (especially diamond, ruby, sapphire, emerald and opal). After this we look at ways in which the colour or general appearance of a gemstone may be altered and how such alterations can be detected.

Finally we look at the rarer artificial materials which have been used as faceted or carved gemstones. Many of these are very rare and collectible, whether or not they may be mistaken for better-known species. While we do not examine their research or industrial applications, their properties are given.

Since so many materials are described in the gemmological literature (there is much more than is commonly thought), the chapters dealing with individual species conclude with a selection from laboratory reports: in this way the 'one-off' stone, which may never find itself the subject of a full paper, gets a mention. Should the reader be faced with a puzzling specimen, it will be possible to look at simple tests (the first section of a chapter) and then turn straight to the report section. Only if nothing is found by that stage will the more complicated test sections need to be consulted.

A critical bibliography, a glossary and an index complete the book.

Readers will find that information is sometimes repeated: I have allowed repetition for the sake of 'if you miss something important in the diamond chapter you may find it in the gem-testing or glass chapter'.

Chapter 3

Further reading

Introduction

The literature on gemstones is much wider than is commonly supposed. While many of the monographs are simply outdated and others fanciful (though pleasing to read), journal papers are written at graduate level and can usually be relied upon – at least until fresh evidence arises! Books on man-made gemstones have been published for centuries but we do not usually call on classical authors to solve present-day problems though they did what they could with whatever knowledge or experience was at hand.

If we are to rely on monographs alone to help us identify man-made and altered gemstones, we are likely to slip up sooner or later since, for obvious reasons, monographs can be published only every few years while a journal article should appear much more quickly, a note in a current awareness newsletter quicker still and information on the Internet almost before the material is around!

In every descriptive study, books repeat each other to some extent, and this is unavoidable. For the student the best course is to keep up with the latest monograph, and supplement its contents with careful reading of the major journals. They are easy to identify: reviews in the *Journal of Gemmology* are the most comprehensive, and readers who need to take a long look at the subject of synthetic and enhanced gemstones should join the Gemmological Association and Gem Testing Laboratory of Great Britain, thus obtaining the benefits of meetings and discussions, with the *Journal of Gemmology* and *Gem and Jewellery News* as vital adjuncts. American readers can subscribe to *Gems & Gemology.* Better still, all readers should subscribe to both: this may seem 'over the top', but if there is a possibility of money being lost on a misidentification, you should take both journals.

Journals

The *Journal of Gemmology* is published by the Gemmological Association and Gem Testing Laboratory of Great Britain, 27 Greville Street, London EC1N 8SU, UK. (telephone (44) 171 404 3334, fax (44) 171 404 8843. The journal carries by far the most comprehensive section of gemmological abstracts and reviews in the world. It appears quarterly.

Gems & Gemology is published quarterly by the Gemological Institute of America, 5355 Armada Drive, Carlsbad CA 92008 USA (telephone (subscriptions) (1) 800 421–7250 ext. 201, fax (1) 310 453–4478; note that 800 numbers are usually obtainable directly only from telephones in the USA and Canada). At the time of writing, only one e-mail address is published in *Gems & Gemology* and this is for editorial queries: akeller@gia.org (Alice Keller is the editor).

The *Australian Gemmologist* is published quarterly by the Gemmological Association of Australia, PO Box 477, Albany Creek, Queensland 4035, Australia (telephone and fax (61) 7 264 6854).

A comparable journal in German is *Gemmologie: Zeitschrift der Deutschen Gemmologischen Gesellschaft*, published quarterly by the Deutsche Gemmologische Gesellschaft, PO Box 12 22 60, Idar-Oberstein, D-55714 Germany (telephone (49) 06781 43011, fax (49) 06781 41616). Recent issues are nearly bilingual German and English and there are English language abstracts for all-German papers.

Readers with access to these journals will have all the information that they need to keep up with most of the productions appearing on the market. Until there is a gemstone bulletin board on the Internet (with all dealers connected) we have to keep an eye open for the small current awareness newsletters which circulate faster than the main journals. One pair is the *Gemmological Newsletter* and the *Synthetic Crystals Newsletter*, both produced by the present writer, from 7 Hillingdon Avenue, Sevenoaks, Kent TN13 3RB, UK. The *Gemmological Newsletter* is entering its 26th year, and the *Synthetic Crystals Newsletter* is not far behind. Both are single sheets and appear 30 and 10 times, respectively, in the academic year.

In Germany, Elisabeth Strack produces her *Strack-Kurier* (in German), which is available from Gemmologische Institut Hamburg, Gerhofstrasse 19, 20354 Hamburg, Germany (postal address Postfach 305287, 20316 Hamburg, telephone (49) 040 352011).

A number of new products appear in the *Lapidary Journal*, published monthly and available from PO Box 80937, San Diego, CA 92138–0937, USA: it carries a large number of useful trade advertisements. The *Jewelers' Circular–Keystone* appears monthly from the Chilton Company, Chilton Way, Radnor, PA 19089, USA: it carries features on different stones and is useful for its topicality.

Crystal growth journals

Since crystal growth has so many applications, accounts of processes and substances can be found over a wide range of journals, all of which are well beyond what you need for recognizing artificial gemstones. For those with a persisting interest, the specifically crystal growth journal is the *Journal of Crystal Growth.* The *Materials Research Bulletin* and Russian cover-to-cover translations under different names also cover the field extensively. They will be found only in major university and national libraries. Do not even think about subscribing! The *Journal of Gemmology*, alone among the gemmological journals, abstracts (makes précis) widely from journals of the related sciences and if you consult the abstracts section you will get a fair idea of what is going on.

Crystal Growth: A Guide to the Literature (1988), by Michael O'Donoghue and published by The British Library, London, is the only guide through some of the maze.

Literature guides

The Literature of Gemstones (1986), by Michael O'Donoghue and published by The British Library, will give you a start to the more specifically gemmological literature. The best general bibliography, covering older as well as modern works, is in the monograph

Gemstones (1988) by Michael O'Donoghue and published by Chapman and Hall, London.

Mineralogical Abstracts, available on CD-ROM and as hard copy, gives abstracts of all papers of interest to gemmologists: the section 'Experimental mineralogy' includes abstracts of relevant papers dealing with new materials with ornamental possibilities: there is also a gemstones section. The publisher is the Mineralogical Society, 41 Queen's Gate, London SW7 5HR, UK.

Monographs on synthetic and enhanced gemstones

The only book specifically aimed at *testing* synthetic and enhanced stones is *Identifying Man-Made Gemstones* (1983), by Michael O'Donoghue and published by NAG Press, London. The best history and description of gemstone synthesis is *Gems Made by Man* (1980), by Kurt Nassau and published by Chilton, Radnor, Pennsylvania, USA. Nassau has also produced the only monograph dealing with the techniques of gemstone enhancement, in *Gemstone Enhancement* (second edition, 1994) published by Butterworth, Oxford. In the gem books series also published by Butterworth the materials quartz, corundum, beryl, amber, jet, topaz, pearl and garnet have their own monographs, in which a good deal on synthetic versions, where appropriate, is included.

The books by Renée Newman, published by International Jewelry Publications, PO Box 13384, Los Angeles, CA 90013–0384, USA, deal with major gemstones in jewellery and with their synthetic counterparts in a way appropriate for counter testing and succeed very well.

While details of older books can be found in the various bibliographies, *Die kunstlichen Edelsteine* (1926) by H. Michel, published by Wilhelm Diebener, Leipzig, gives an excellent overview of early gemstone manufacturing methods and for those with a particular interest in the patenting of gem crystal growth methods (the art is in what you can leave out!), *Synthetic Gem and Allied Crystal Manufacture* (1973), by D. Macinnes, and *Synthetic Gems Production* (1980; to some extent an update), by L. Yaverbaum, both published by Noyes Data Corporation, Park Ridge, New Jersey, USA, give details of many obscure patents which would otherwise be hard to track down.

For simple gemmological study, *Gemmology*, by Peter Read (Butterworth, 1991), outlines the theory behind the testing of gemstones, and is particularly good on gem-testing instruments, the author's speciality.

No one book can be said to cover all the ground, but the bibliography in one book (and the literature guides cited above) will lead you to more information elsewhere: a good deal of it is (or seems) contradictory, but this is the way of textbooks!

Chapter 4

The growth of gem quality crystals

Crystal growth is complicated and unpredictable. The grower will be asked to manufacture a crystalline substance with particular properties, perhaps optical or electrical or, in the case of gemstones, coloured or colourless, transparent, translucent or opaque. Gemmologists usually know little about the problems facing and the techniques used by the crystal grower, even though a great deal of money may hang upon correct identification of a synthetic ruby or emerald.

A brief overview of crystal growth shows that almost all crystals grown today are unsuitable for ornamental use because they are too small. While there have been periods when a number of hard, relatively large transparent substances have been grown, in general the greatest efforts have been towards growth of thin crystalline layers of one material on a substrate of another ('butter on bread'). None the less many materials of great beauty and interest are in existence, and there are some major collections of as-grown and fashioned synthetic substances: a visit to St Petersburg and Moscow will locate at least two of them which I carefully examined in 1975.

For ornamental use a crystal must be able to withstand possible damage caused in the fashioning process and in normal conditions of wear. The crystal must look attractive and be large enough to be seen. Very soft substances and those with an easy cleavage can be ruled out as can those with a chemical composition likely to be harmful to the wearer. Transparent colourless crystals should have a high dispersion and as low a birefringence as possible: if dopants (chemical elements not forming part of the regular composition of the grown crystal) can be added to give a range of bright and unusual colours so much the better.

What is written above is most applicable to the growth of 'new' substances with no natural counterparts. Since ruby, sapphire, emerald, quartz and now diamond can be grown with some skill, some at least of the world's crystal-growing efforts will be directed towards such profitable items. Next comes the question of cheapness: there are many methods of crystal growth, and those which give the largest crystals for the smallest cost will be preferred. This is why about 90 per cent of corundum and spinel crystals are grown by the simple and cheap Verneuil flame-fusion method.

Details of the thermodynamics involved in crystal growth can be found in the large (and expensive) crystal growth literature, a guide to which (O'Donoghue, 1988) can be found in the bibliography. We need to consider what we need from a gem-quality crystal, apart from the considerations already mentioned above. Is there an element of deceit in the process and if there is, does it matter? How much effort is put into making the grown crystals 'look natural'? Without definite knowledge and in the absence of comment from gem crystal growers we cannot say for sure! My feeling is that the commercial crystal growers do not trouble about whether or not their products need to be tested – they are in business to sell them.

The chemical composition of a mineral or gemstone determines the method to be used for growth. Among the gem minerals oxides and silicates occur most often, elements (apart from diamond) hardly occur and many sulphates are water-soluble! There are few gem sulphides, arsenates, carbonates and phosphates which is just as well since they are not always very easy to grow in large sizes.

Looking at the oxides first, the simple combination of a metal with oxygen (corundum, quartz, spinel, rutile) makes it possible for crystals to be grown from a starting material (feed powder) consisting of the necessary elements with whatever is needed for colouring. The feed powder can be fused by a flame hot enough to melt the powder, droplets of which collect on a ceramic pedestal to form a single crystal known as a boule from its characteristic shape. Since corundum has a melting point of 2050°C it is clear that crucibles able to sustain the melt would be hard to find! They do exist, but are very expensive, as we shall see later. The way in which the boules form by slow accretion of rapidly cooling material in successive layers causes many of them to show curved lines resembling the grooves on a vinyl record, the lines long held to be the best way of identifying corundum grown by this method. It is true that many colours of corundum grown this way do show the lines but they are not so easy to see as the textbooks suggest: in colourless and yellow sapphires they can be seen only under special conditions which we shall meet in due course – they are easiest to see in the oddly slate-to-purple coloured specimens doped with vanadium and intended to simulate alexandrite. Such stones can be found in quite large sizes, and it is probably easier to see the lines when the stone is sufficiently large to manipulate under magnification without having it fly from the microscope!

Without the need for an expensive crucible and with the hot flame provided by cheap sources of oxygen and hydrogen, and with the possibility of many furnaces being supervised by one person and no serious limit to boule size, it is not surprising that the flame-fusion method, usually known as the Verneuil method, after A.V.L. Verneuil, the French scientist who developed the technique in the nineteenth and early twentieth centuries, should be the choice for the large-scale growth of gem oxide crystals (Figure 4.1).

The use of gas to provide two flames causes gas bubbles to be incorporated into the finished boules. They show as large, well-rounded bubbles with thick, bold edges, these showing that there is a considerable difference in the refractive index of the contents of the bubble and that of the corundum host. If you cannot see the curved growth lines or curved distribution of colour (best viewed when the specimen is immersed in a liquid of similar refractive index) the gas bubbles will give a distinctive clue to the Verneuil product. A further clue, applicable to coloured varieties of corundum but not to spinel, is that both the finished single-crystal boule and natural corundum crystals have their long axis (known as the *c*-axis) in the same direction. As it happens, this axis is a direction of single refraction in a material that is doubly refractive in all other directions: this is not all, because a stone cut with its table facet parallel to the axis will show pleochroism through it. This is unlikely to happen with natural corundum crystals since stones cut this way will not be thick enough. The synthetic corundum shows the effect because the lapidary can get more stones from the boule by placing the table parallel to the long direction. This applies only to corundum grown by the Verneuil method.

The whole Verneuil process from boule growth to the sale of faceted stones is mass production. Stones may be faceted mechanically and fast and may show characteristic surface markings resembling wavelets: the name ‘fire-marks’ has been given to them. A

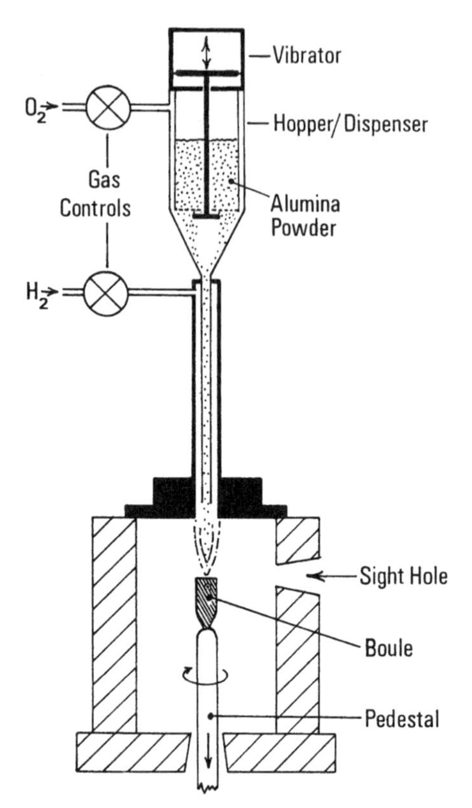

Figure 4.1 Corundum growth by the Verneuil method. (After P.G. Read)

combination of all the effects described is quite enough to diagnose a Verneuil corundum.

Spinel is grown by the same method but here the composition is not quite the same as that of natural spinel. Extra alumina has to be added to achieve satisfactory boule growth: a typical composition for most Verneuil spinel would be $Mg2\frac{1}{2}Al_2O_4$. This has the effect of increasing the specific gravity from 3.60 to 3.64 and the refractive index from 1.718 to 1.728. Stones, while not usually showing so many useful inclusions as Verneuil corundum, contain oddly shaped bubbles and, between crossed polars, show a striped effect as the specimen is rotated rather than remaining dark as cubic minerals usually do. Remember that synthetic spinel is manufactured to imitate other gem species.

Other species are made by flame-fusion, but when crystal growers want to grow silicates or oxides of higher quality (more closely resembling the natural stones) other, slower methods are used with a view to attaining greater perfection. Verneuil crystals, despite the inclusions and colour distribution, tend to be glassy, while natural specimens, with light scattered from inclusions, do not have this rather bare appearance. Crystals grown by slower methods incorporate different types of inclusion which 'look natural' and may cause difficulties for gemmologists.

Slower growth needs a crucible to hold the melt which, in the case of corundum, is at 2050°C. Since the crucible must not contain elements which will form unwanted compounds with the feed material, precious metals such as platinum or iridium are used (at great expense), and since growth of a crystal large enough to provide a 2–3 ct stone may take up to 1 year of slow, controlled cooling, it is easy to see why these specimens cost far more than Verneuil ones which can be made in hours.

Top-quality rubies are grown by the flux-melt technique, and are usually called flux rubies. The same method is used to grow multicomponent silicates which cannot be grown by the Verneuil method, so emerald, as beryllium aluminium silicate (four components) is also flux-grown. The word 'flux' means 'solvent' in this context: the starting materials (ingredients) for ruby or emerald are first dissolved in a compound (flux) which then melts at a lower temperature than would otherwise have been necessary. Slow cooling is the vital part of the process, though choice of flux affects the form (shape) of the finished crystals. For gemstones a fairly blocky crystal is needed rather than a needle-shaped one! As well as ruby and emerald, alexandrite, spinel and some of the synthetic garnets are grown in this way. Properties are the same as those of the natural material: though some emeralds have a slightly lower specific gravity and refractive index, the readings are not diagnostic.

This method produces gem mineral crystals which can cause problems as they 'look natural', and since most gemmologists do not see good-quality gem crystals they may not recognize the artificial product. Crystal groups can be grown, and these do not occur with natural ruby or emerald. With faceted stones, considering emerald first, the grower can introduce chromium and exclude iron, so that stones when viewed through the Chelsea filter show a very bright red, which to some extent separates them from most natural emeralds where iron is usually present to a greater or lesser amount. Much more obvious, when they have become familiar, are twisted veils or smoke-like inclusions of undigested flux. Most of all, natural emerald shows a number of natural mineral inclusions which are absent from flux-grown material.

Rubies also show flux veils and lack mineral inclusions: both rubies and emeralds may show angular metallic fragments from crucible material. Growth usually takes place on a cut seed crystal, and trace of this may remain – though normally the finished stone would be cut from parts of the crystal well away from the seed. Sometimes clouds of flux take a roughly hexagonal pattern: this is often seen in the rare Zerfass emerald.

Crystal growers have developed several techniques for flux growth but identifying the resulting gemstones involves the same tests. Alexandrite will also show flux particles, which seem less common in the synthetic garnets (most of which, to be fair, are grown by crystal pulling, described below).

Any ruby, emerald or alexandrite with no visible natural inclusions, with wisps or veils of flux particles, or occurring in multicrystal groups, should be carefully tested.

While many species can be grown by the flux-melt method, large-scale production of crystals for industrial use is more easily carried out by other techniques. Ruby and emerald do not need to be made in such large amounts, but quartz, so valuable in its widespread electronic applications, can be grown in very large sizes.

Growth of quartz and some emerald is by the hydrothermal process, in which the starting materials are melted, with water, in a sealed pressure vessel (autoclave). Growth is aided by the addition of a mineralizing compound, and takes place on the surfaces of prepared seeds. Since quartz crystals are grown for electronic applications rather than as ornament, the cutting of seeds at appropriate angles is a matter of great importance. Pressures in the vessel are high, and the name 'bomb' has been quite appropriate on occasion.

Quartz crystals have to be inclusion-free to be satisfactorily used in watches, so that the flux-melt method of growth is ruled out: flux inclusions could not be tolerated. For the

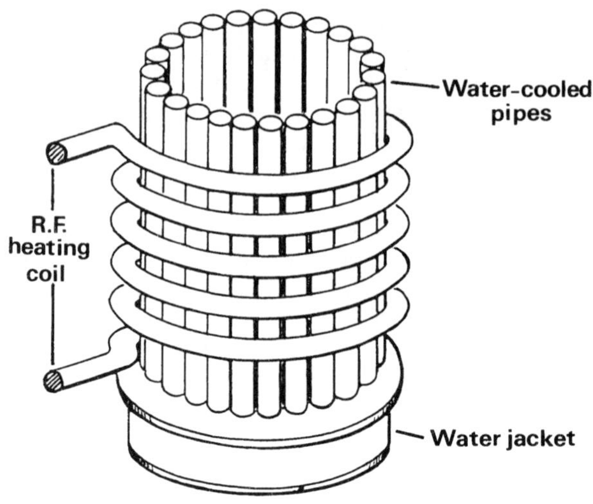

Figure 4.2 Skull melting apparatus

gemmologist, choice of the hydrothermal method means that there are very few, if any, inclusions in synthetic rock crystal, citrine or amethyst, which in any case have the same properties as the natural material. As always, the absence of natural inclusions should suggest an artificial product.

Hydrothermally grown emerald may contain pointed growth tubes and other distinguishing features described in the emerald chapter (Chapter 9): ruby has been manufactured mainly for research purposes and no hydrothermal ruby appears to be on the market at the time of writing. I have seen one example of a ruby crystal in which the colourless seed is visible. As platinum wires protrude from the crystal it is unlikely to be mistaken for natural ruby!

The development of the laser was responsible for increased research into and growth of inclusion-free ruby crystals. With a melting point of 2050°C the hydrothermal method is not completely suitable, and the flux method would produce unacceptable inclusions. Crystal pulling, often known as the Czochralski method, gives ruby rods of high purity (the method is also used for a variety of other substances, some occasionally appearing as faceted stones). The only inclusions may be elongated bubbles, but these are not common and would be ruled out for laser crystals.

The method involves lowering a seed crystal to the surface of a melt in a crucible, followed by its slow withdrawal upwards. The melt is taken up with the seed to form a cylinder or rod. Control of the different parameters is difficult. Rejects from the laser crystal programme often finish on the lapidary's bench.

A number of other methods are used for gem crystal growth, but since none of the products contain natural inclusions there is no particular need to distinguish between them: the reader is referred to the extensive crystal growth literature.

Cubic zirconia has a melting point of 2750°C, and a crucible cannot be used for growth. Instead, the technique of skull melting is used (Figure 4.2). Here the crystal grows from a melt begun within a block of its own powder, the outside of which is kept cool by circulating water. Heating by radio-frequency induction is assisted by placing pieces of zirconium metal within the powder. On cooling, columnar crystals are harvested. As with all clear colourless materials, doping to give a range of colours is a matter of routine. There are no recognizable inclusions and the reflectivity meter is the best method of testing – if you need to know if a specimen is diamond or not. Identification of cubic zirconia from a group of miscellaneous colourless faceted stones can be difficult.

The unique properties of diamond make its successful synthesis difficult, and the literature is full of descriptions of attempts made. Since diamond growth is so complicated a process, the reader is referred to specialist books on it: these can be found in major scientific libraries.

Chapter 5

Gem testing

As you go through this book you will find many terms which need explanation: specific gravity, refractive index, absorption spectrum, crossed polars, inclusions and many others all form part of the vocabulary of gem testing. The book is not intended as a textbook for gemmology students but as a guide to artificial products masquerading as natural gemstones: however, you will need to know something about simple gem testing even if your specimen has to be passed up the line to a laboratory or professional gemmologist.

Details of the various gem-testing instruments are exhaustively covered by a number of textbooks, and without going too deeply into how they work, we shall look at what you do when one of them seems an appropriate source for identification. Much can be done with a 10× lens, but ironically, as so often happens, it comes into its own only when you know enough to be able to use it profitably: by then you will probably know that another instrument will give you an answer which may be diagnostic. By 'diagnostic' we mean a result from a test which tells you that your specimen must be a particular species and no other. When you get such a result there is no need for confirmatory testing by another means. A good example can be seen in the valuable green demantoid garnet, which contains wispy chrysotile inclusions resembling horsetails: such inclusions are seen in no other green stone and they are also quite easy to find.

Apart from glass and opal, all gemstones are crystalline. This affects gem testing in several important ways. Crystals are grouped into seven systems, in all but one of which light entering the crystal is split into two rays which pass through at different velocities – such specimens are *anisotropic*. Both rays have their own refractive index (the ratio between their velocity in the stone and the velocity of light outside). The crystal systems vary in their symmetry: crystalline symmetry is three-dimensional, and differences between one crystal and another depend upon the ways in which atoms of their constituent elements are packed together. Crystals with the highest crystal symmetry belong to the cubic system and as their atomic packing is 'the same all round', all rays of light traversing the stone do so at the same velocity and there is only one refractive index – they are *isotropic*. Diamond and the natural and synthetic garnets belong to the cubic crystal system.

Glass and opal are not crystalline (this *amorphous* state is rare in nature) because they form too quickly for their constituent atoms to take on a regular structure. Amorphous bodies have no directional properties as crystals usually do, although they do have an atomic structure, of course. They can show no bifrefringence (the arithmetical difference between the two refractive indices present in most crystals) and no pleochroism (showing a different colour or shade of colour in different directions). Nor can amorphous materials

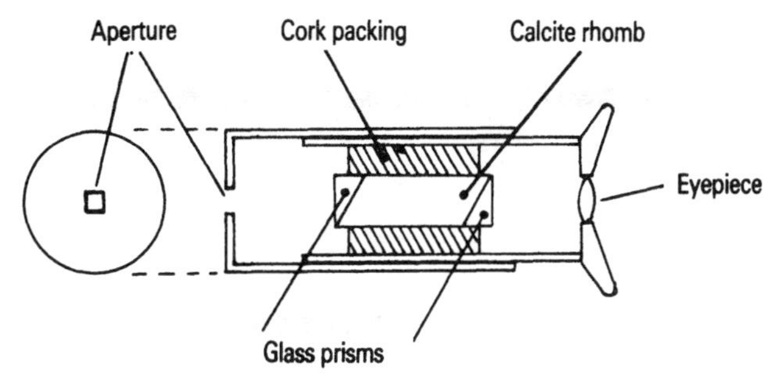

Figure 5.1 The dichroscope

show directional hardness – but hardness is not usually a major feature in gem testing.

The form taken by the atomic structure of solid substances depends on the chemical elements present, so that if you could find an instrument which would tell you which ones were there, you would have a self-contained gem-testing laboratory! Sadly, nothing is quite so simple, but the spectroscope, one of the most powerful tools in chemical analysis, goes some way towards achieving it. To a chemist, 'spectroscope' means grey boxes, under computer control, producing pictures of peaks and troughs (and interpreting them as well) over a large part of the measurable electromagnetic spectrum. To the gemmologist the 'direct-vision' spectroscope operates only in the visible light region, so that you can see some of the elements present. It operates on crystalline and amorphous bodies alike.

Coloured crystalline substances usually show pleochroism – difference of colour or shade with change of viewing direction. Such effects are usually subtle but sometimes they can be spectacular. The dichroscope (the terms 'dichroism' and 'pleochroism' are used quite loosely) presents, via a highly birefringent crystal of calcite, a doubled image of a small rectangular window at the opposite end of a small tube from the eyepiece (Figure 5.1). A pleochroic specimen, correctly oriented, will show the two colours side-by-side: some crystals may show three distinct colours, but for this the stone needs to be repositioned and another observation made. The presence of two colours in the windows rules out glass and cubic minerals and, when the effect is strong, helps to identify the fairly few species where this may be expected. Gemmologists take care to look in different directions, since even in birefringent crystals there is always one (sometimes two) directions where they behave like cubic or amorphous materials. These *optic axes* have to be reckoned with but if you remember always to alter the position of the specimen you should not be caught out!

The difference between singly refractive (cubic crystals and amorphous materials) is also well shown when polarized light is used. Unpolarized light has vibration directions in all directions at right angles to its direction of travel (like an axle with an infinite number of spokes); polarized light 'has only one spoke', and can be put to a variety of uses. In the polariscope two pieces of Polaroid (as in sun-glasses) are positioned above one another with a space in between for the specimen and a source of light beneath. The two Polaroids are so placed that all vibration directions from the light are cancelled out and when the light is on you can see nothing through the upper Polaroid: this is called crossed polars. When a specimen is placed between the two some light may pass through to the eye as the stone is rotated: the patterns of light and darkness are characteristic for specimens with different structures, so that cubic minerals remain

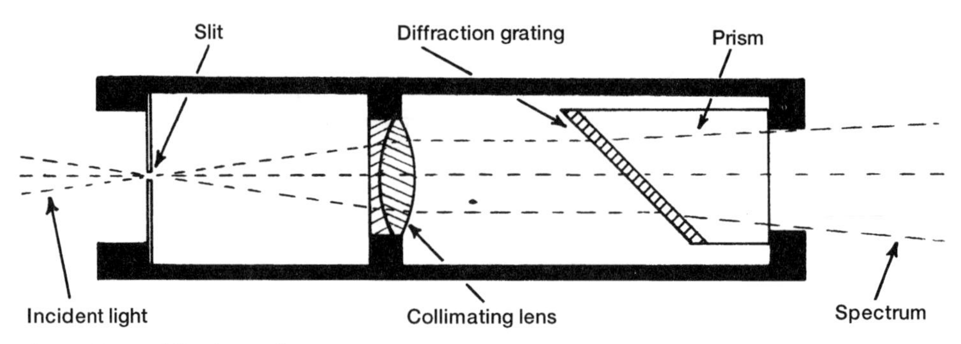

Figure 5.2 The diffraction grating spectroscope

dark throughout a complete rotation while crystals of the other six systems vary from light to dark four times in a complete rotation. Nothing is ever perfect, and it so happens that many of the cubic gem minerals do not remain dark as they should but give a stripy pattern instead. The stripes, resembling those on a tabby cat, have led to the term 'tabby extinction' for this effect, among gemmologists at least. When seen it is highly suggestive of synthetic spinel (a member of the cubic crystal system) but it can be seen in other species too.

This test does, therefore, give some distinction between a natural and a synthetic product, but on the whole, since natural and synthetic gemstones are the same compositionally, their responses to most tests will be the same. Only the spectroscope and microscope will be vital in this area of testing.

The hand (direct-vision) spectroscope examines the specimen in a strong white light which may either pass through or be reflected from it (Figure 5.2). The observer will see a ribbon of spectrum colours, crossed vertically by dark (sometimes light) lines or bands, the pattern of which indicates a particular chemical element present in the specimen. Dark bands are known as absorption bands ('lines' and 'bands' are used loosely), and coloured ones are emission bands. Note that all emission bands emanating from the specimen will

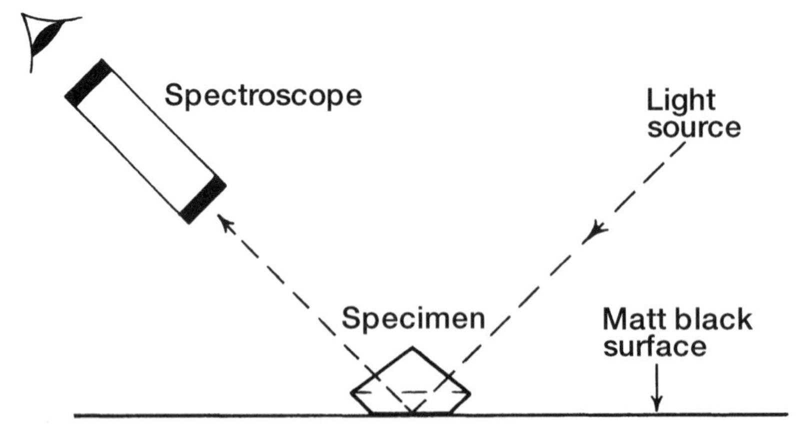

Figure 5.3 Good spectroscopic technique

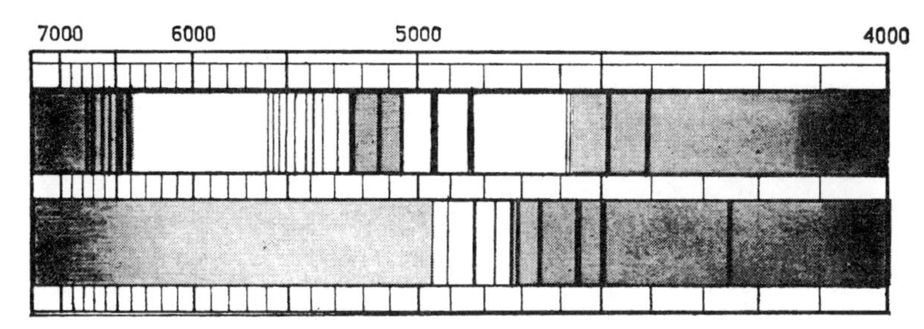

Figure 5.4 Rare earth absorption spectra

be red: any other colours will arise from the room lighting. Since the eye takes time to adapt to observation, the room should be darkened before work begins so that faint absorptions can be seen more easily. Figure 5.3 shows how to increase the path length of light through a specimen and thus maximize the depth of coloration of the transmitted light.

A series of very fine lines extending over most of the visible spectrum will indicate a rare earth element, and show that the specimen has been doped with a rare earth for colouring purposes (Figure 5.4): such a specimen will be artificial. The absorption spectra shown by a variety of natural and synthetic gemstones are shown in detail in all gemmology textbooks. Bands are referred to in nanometres (nm) and the visible spectrum extends from about 700 nm (red light) to 400 nm (violet light).

The microscope (Figure 5.5) is the 'desert island' gem-testing instrument! A suitable model can give a range of magnifications (though 60× at most would cover almost all routine testing, and much can be done at 20–30×), and can also be adopted to measure the refractive index, display pleochroism, serve as a light transmission agent for spectroscopic observations and act as a polariscope! How these tests are accomplished can be found in the gemmological textbooks. Here we need to look at the magnifying powers of the microscope: it is easier to use than a 10× lens since the specimen can be held still and moved easily into different positions. Most microscopes provide bright-field and dark-field illumination, the latter lighting the specimen from the side so that inclusions can be seen brightly against a dark background.

By 'inclusion' we mean anything seen inside a specimen but we can also observe details of the surface, of faceting and of damage. Inclusions are a major field of gemstone study, probably the most important one for the gemmologist who does not need refined chemical analyses but does need to be able to recognize the internal furniture of natural and synthetic products.

Inclusions may be solid (mineral), liquid or gaseous and the absence of solid inclusions is the most characteristic feature of man-made gemstones. While all the inclusions you may expect to find cannot be described in such a book as this, we can at least pin down some of the fanciful terms beloved of some older gemmologists and which do not mean what they appear to mean: I have tried to explain these terms as they arise. Clearly, to be able to recognize solid inclusions (or their absence – not so ridiculous as it sounds) you have to look at as many as possible in specimens whose nature has already been established. Liquid inclusions are harder to see and hard to describe: gaseous ones either stand alone (the specimen is glass or a Verneuil synthetic) or form one phase of a multiphase inclusion (usually though not always in a natural specimen). They may also turn up in plastic or glass fracture fillings.

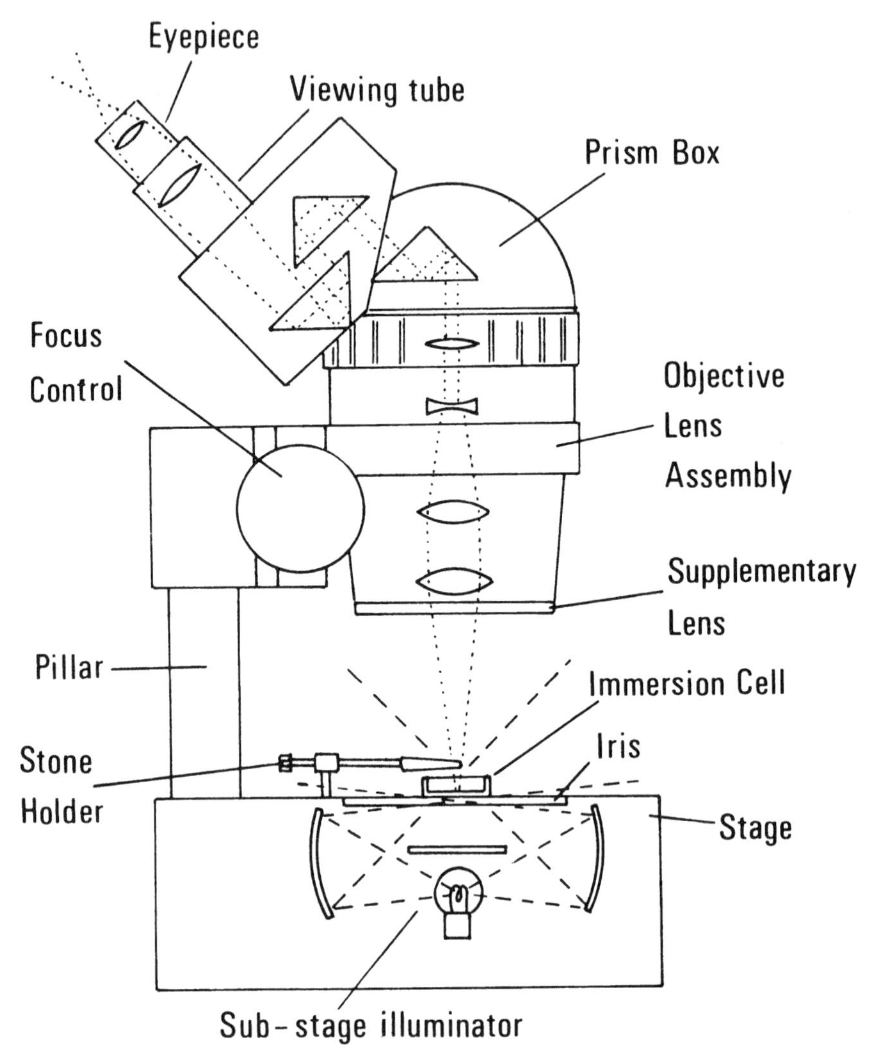

Figure 5.5 Stereoscopic binocular microscopic. (After P.G. Read)

An old friend of mine says 'An Englishman needs time', and though he had a quite different context in mind, time is certainly needed for the gemmologist armed with a microscope. The lighting and the specimen need constantly to be moved until the best effects are found. Then books of photographs need to be consulted in the hope that some similar feature will be there. Most gem testing involves the observation and evaluation of several different features before a verdict can confidently be given.

Gemmology students of past generations will wonder why the refractometer was not considered first among this survey of gem-testing instruments and techniques. The instruments already discussed operate equally well on mounted stones as on loose specimens: on the refractometer most specimens need to be free from their setting, as parts of it unfailingly get in the way and prevent the stone lying flat, as it needs to do.

More importantly, the liquids used to make optical contact between a specimen and the glass of the instrument have been found to be hazardous to health, and it is surprising that the dangers have not been publicized before. At the time of writing, a substitute liquid for the one long in use has been devised, but I am sure that in time this also will be found unacceptable. Whatever the consequences of further investigation it would be better all round if gemmologists could improve their skills with the microscope rather than turning to the refractometer for most of their testing. To some extent gemmology has limped behind other investigative sciences, perhaps because in the past most practitioners have not been graduates: persistence with dangerous chemicals is a rather negative result of a certain stolidity characteristic of past years.

None the less, there are times when the refractometer has to be called up for service: when birefringence needs to be measured rather than merely estimated and when a specimen shows no useful pattern of inclusions. Details of its operation are found in many textbooks: all that we need to say here is that a stone is placed table facet down on a glass window, optical contact being maintained by the intervention between the two of a drop of the liquid aforementioned. The refractive index can be read from a scale visible through the eyepiece, or combined with a side knob which operates a shutter whose bottom edge is made to coincide with one or another of the possible refractive index readings, seen as a more or less sharp division between the upper dark part of the visible field and the lower bright part. If monochromatic sodium yellow light is used (for which the instrument is calibrated), the division will be quite sharp; if room light or daylight is used, a band of spectrum colours will be seen at the refractive index position and this is often easier to see than a vague division between light and dark.

Stones (the majority) which have two or more refractive indices may show them both: as the stone is rotated their maximum and minimum positions should be noted; simple subtraction gives the birefringence (double refraction or DR), a very important property. Some stones have a refractive index too high to be read, and are said to show negative readings. Stones which have optic axes will not show birefringence if one of the axes coincides with the axis running from front to back of the refractometer: rotation of the specimen is always necessary so that this possibility does not become a reason for misidentification.

Observation of the shadow-edges dividing light from dark can provide a lot more information than merely refractive index or birefringence: consult a textbook for further details. None the less, the liquid does pose a problem, and conditions of use should be carefully monitored.

Specific gravity is defined as the weight of a substance compared to the weight of an equal amount of pure water at 4°C (when water is at its densest). If A is the weight in air and W is the weight in water, then SG = $A/(A-W)$, the result of this calculation being multiplied by C, a constant denoting the SG of whatever liquid is actually used for the determination: if water, then C=1 and no further calculation is needed. If another liquid is used then its SG would be C. Laboratory determinations would take place at the designated temperature, and for determining the SG of quite large specimens such as gemstones distilled water is usually quite adequate. For such tests the specimens need to be unmounted (though some necklaces can be tested as the string makes little difference) and fairly large to avoid system error and difficulty of handling.

The hydrostatic method of SG determination in which successive weighings take place in air and when totally immersed in water is accurate but slow and prone to error. It is described in all gemmology textbooks. A quicker method of determination is to observe the rate of sinking in a ('heavy') liquid of known SG: specimens float, sink slowly or fast, or remain suspended at a position to which they are pushed by a glass rod – at which position there is a close match between the specimen and liquid. Surface tension needs to be taken into account in both methods, and some of the liquids commonly used are hazardous to health.

As most synthetic stones have the same SG as their natural counterparts, an SG test is less likely to be useful than one with the microscope or spectroscope. Should you ever need to test a lot of small stones at once, the 'heavy liquid' test might sound a good choice – but retrieving them would take a long time and as the liquids are quite expensive they should not be poured away in desperation (through a filter paper, of course)!

The use of ultra-violet radiation can be quite a useful back-up test for gemstones even if it rarely produces a firm diagnosis. Long-wave (LWUV) and short-wave ultra-violet (SWUV) radiations are produced with the aid of special filters which pass some visible light (leading to the exclusive band of those who claim to 'see' UV light). Gemstones containing chromium and which are iron-free will often give a notably strong red response under LWUV and, since iron-poor stones would be a rarity in nature, such a response would certainly suggest a synthetic stone. Colourless synthetic spinel, often used to replace small missing diamonds, fluoresce a bright sky-blue under SWUV (some synthetic diamonds show a similar response!). There are many other useful examples in this and other books.

When the UV light is switched off and the specimen continues to glow, albeit for a short time, the effect is known as phosphorescence. A colourless stone fluorescing blue and phosphorescing yellow, at the time of writing, has to be diamond (thus a diagnostic test(s)). X-rays act in the same way but are not quite so useful: synthetic Verneuil rubies show a strong red phosphorescence.

Some synthetic materials show unusual or unexpected fluorescent effects as the result of doping, and any stone whose colour (or lack of colour) seems unusual could well be tested under UV radiation.

Larger than life

We have mentioned the 10× lens casually as we have gone along but make sure that when you get one you make its use easy for yourself. The best way to do this is to ensure that light falls on to your specimen and not into your eyes: it is interesting to see how beginners 'stand between themselves and the light'. The lens will enable you to see examples of how gemstones belonging to the non-cubic crystal systems will show inclusions and back facet edges (viewed through the top (table) facet) apparently doubled. This is due to the splitting up of the incident light into two rays as it traverses the specimen.

As we know, the big brother of the 10× lens is, of course, the microscope, which will do whatever the lens will do but is larger and more expensive.

What do you do when all else fails?

There are times when a stone is too small for the refractometer or for SG testing, colourless (no absorption spectrum – or one is unlikely), or is large enough to sit on the refractometer but has too high a refractive index for the instrument to give a reading.

Perhaps the stone is mounted with a closed setting! Perhaps it is one small diamond-like stone among many others in a large piece of jewellery.

Not so long ago the answer to the question would be 'send it to a laboratory', but this can be expensive. The reflectivity meter gets round the problem for some species: testing the surface reflectivity of a faceted stone by using an infra-red beam will distinguish a diamond from any of its simulants, and some reflectivity meters will identify the simulants too.

When we learn that diamond has unique powers of conducting heat we might wonder whether this property might also be made the basis of a gem-testing instrument. In fact the *thermal conductivity tester* has been around for many years now: with its small copper-tipped probe it can reach very small stones in deep settings in a large piece of jewellery so that a rogue glass or colourless synthetic spinel shows up at once. Usually the thermal conductivity tester only diagnoses diamond, not stating what the rogue stone is.

The thermal reaction tester is mentioned in a number of the reports. Older books call this the 'hotpoint' as do all gemmologists. An electrically heated probe is brought near to, say, a suspected plastic (carefully – the customer may value the plastic), whereupon it will arouse a characteristic nose-tingling smell (or a rubber smell from the vulcanite imitation of jet).

Chapter 6

The major natural gemstones

Before we look at synthetic and colour-enhanced stones we must know something about the 'real' specimens so that we can distinguish them when testing. While we cannot say too often that a synthetic stone *is* ruby, emerald, alexandrite in every respect (chemically and physically) but for its being man-made, there are still features which distinguish natural from artificial products: these features reflect the growth processes of nature or of man.

In this account, repeated in one form or another in many places through the book, we take a brief look at those properties used in routine gem testing. Most attention is paid to what can be seen through the lens or microscope since this is the theatre from which the diagnosis eventually emerges.

Hardness (H) is not usually a test for faceted stones nor for the often beautiful rough gem material: none the less, there are occasions when it is useful for both the gemmologist and the lapidary to be aware of it. Cleavage, the propensity to break along directions of atomic weakness, found only in crystals, gives useful clues in testing and sometimes disasters for the lapidary. The *specific gravity* (SG), the ratio of the weight of a specimen to the weight of an equal amount of pure water at 4°C (when water is at its densest), is a useful test for unmounted specimens, though tedious and usually unnecessary if your microscope/spectroscope technique is adequate.

The *refractive index* (RI) provides an easier way of identifying a specimen than an SG test and birefringence (*DR, double refraction*) or the lack of it, is equally suggestive. Both measurements relate the velocity of light within the specimen to the velocity of light under normal conditions 'outside'. LWUV and SWUV refer to long- and short-wave ultra-violet radiations (often referred to as 'light' though you cannot see them): again, effects seen under the radiations may give a useful back-up to the commoner tests and occasionally be diagnostic in themselves. The *spectroscope* provides (if you are lucky) an *absorption spectrum* – a ribbon of spectrum colours with distinctive and sometimes diagnostic vertical dark lines or bands (terms loosely used) superimposed on it. We should expect coloured stones to show absorption spectra rather more frequently than colourless ones but beware! Some diamonds which appear to be colourless will show at least one absorption band as you will find in the course of the book. Coloured stones may also show useful *pleochroism* (*dichroism*) – again, terms used loosely: the specimen shows different colours in different directions. While the difference is often quite easily perceived by the eye, the *dichroscope* is a handy small instrument which can be used to make sure you are seeing what you think you are seeing, a problem associated with this effect and deriving from the inability of the eye to distinguish colours adequately, this varying greatly with lighting conditions and with the individual.

These are the properties which are, with some exceptions, generally shared by all gemstones. Gemstones belonging to the cubic crystal system can show no pleochroism or birefringence; nor can glass or opal, which are amorphous, belonging to no crystal system, as their constituent atoms take up no long-range order as in crystalline substances.

These properties and tests are more exhaustively discussed in various gemmology textbooks: in this book we look at them only in so far as we need to know what we are talking about when we come to compare natural with synthetic specimens or with those whose colour has been altered. We have already looked at the instruments with which simple tests can be carried out and get a rudimentary idea of their operation.

The best of all gem-testing instruments is the microscope: while it is not difficult to see inclusions (things inside stones), what is hard is the art of so describing them that others will know what we are talking about. The 'language of inclusions' inclines to the fanciful: while this would not matter if the nouns and adjectives used meant the same to everyone, this is clearly not the case, and when a description needs to be translated from one language to another the meaning can easily be lost. There is, therefore, a good case for learning the simple technical terms which describe what the inclusions actually are, replacing 'silk' with rutile and 'horsetails' with chrysotile. There should be no difficulty, any more than replacing 'pigeon's blood' with 'traffic-light red' when ruby is the subject.

Inclusions can be solid, liquid or, sometimes, gaseous, though when large, well-rounded gas bubbles are seen on their own, rather than forming part of a two- or three-phase inclusion, they are strongly suggestive of natural or artificial glass, or of the crystals grown by the Verneuil flame-fusion technique. By 'two- and three-phase' inclusions, we are referring to the three states of matter, solid, liquid and gaseous. The term 'negative crystals' means a crystal-shaped hollow, sometimes empty, sometimes filled by mineral matter or liquid: the shape of the crystal reflects that of the host mineral. The commonly used term 'feathers' or 'fingerprints' usually refers to flat planes made up of liquid shreds, tubes or patches: when such structures are twisted there is a strong suggestion that their host may be a flux-grown crystal.

While pleochroism is the difference in colour seen with difference in direction of viewing, many natural gemstones show an uneven colour distribution – in fact in many classic stones such as blue sapphire, colour zoning seems to be the rule! Most gemstone dealers and many jewellers will be familiar with the Sri Lankan blue sapphires with just a spot of blue colour in the base (culet) of the stone and which appear entirely blue when viewed through the table (top). Uneven colour distribution is not in itself a sign of either natural or artificial origin, but a stone with no sign of uneven colour certainly needs to be carefully examined. In general, colour is neither a good nor reliable guide either to the origin of a natural gemstone or to whether a specimen is natural or artificial. Perceptions of colour vary from individual to individual, from day to day and from time of day to another time: they may even depend on the type of meal recently eaten! I labour this point because many dealers will affirm that they can 'tell a synthetic by its colour': this may sometimes be true, but the value of this kind of statement only becomes apparent when the number of correct (or incorrect) identifications is large enough to become a statistical sample.

There are a few instances of the faceting style used for a stone being suggestive of artificial origin: a specimen with the scissors cut is very likely to be a Verneuil-type synthetic, as this is a cheaply achieved style of cutting which can easily be carried out under computer control. The lens will suffice for identification here. Likewise, a star stone with a neatly polished flat back, fine even colour and a perfect, well-centred star does rather suggest a Verneuil synthetic: natural star stones are usually rather lumpy (to increase weight), have ill-centred stars and/or indifferent colour.

Having established some of the criteria we shall be using to look at those specimens of unknown nature and origin which may be presented to us in the course of a working day, we can now consider the major gem species as natural specimens: in each case the properties will be given in the same order for ease of reference so that when we come to look at synthetics we shall have something to compare them with.

Diamond

You may be offered diamond rough rather than faceted stones: though in many diamond-producing countries it is illegal to possess diamond rough, the opportunity may arise. Rough diamonds look like washing soda, we have often been told, but not many people know what this looks like! Diamond crystals may show the familiar octahedron (two square section pyramids joined at the base) or may be flat and glassy: not many other natural crystals resemble diamond but flux-grown colourless corundum and spinel both show octahedral form and can be sharp edged – too much, in fact for diamond. Diamond crystals have an unmistakable bright and somewhat greasy lustre, for which the name 'adamantine' is used.

We should not forget that diamond is not always colourless or that faint hint of yellow known as 'Cape' colour. Coloured diamond rough is really rare, and in many years I have seen little outside exhibitions and not often there. However, a fine red or green crystal of yttrium aluminium garnet (YAG) or gadolinium gallium garnet (GGG) can 'look like diamond if you don't know what to look for'.

If you are uncertain about a diamond crystal do not try a hardness test in case you start a cleavage (diamond has four possible directions in which this can happen). An SG test will show diamond reading 3.52.

Before going on to faceted stones, here are the main properties of diamond: H 10; SG 3.52; RI 2.417; dispersion 0.044; variable fluorescence and phosphorescence. Cape stones show a sharp absorption band at 415.5 nm: details of the spectra of coloured and treated stones are given in the main diamond chapter.

It is well worth becoming familiar with the common inclusions of natural diamond. Almost all faceted diamonds will contain some inclusions: all are solid (i.e. mineral), and provide the best means of distinguishing diamond from its imitations. Solid inclusions are most commonly fragments of diamond and their unmistakable bright lustre is easily seen with the lens. Other minerals are olivine, enstatite and diopside, all green, with olivine paler than the others. There is no need to establish what the inclusions are: all you need is to know is which ones establish the identity of the host. Crystals of red pyrope–almandine garnet are fairly common, and some stones have been found to contain yellow needles of rutile. Diamond contains no liquid inclusions: while their absence indicates formation under conditions of great heat and pressure, it is useful to the gemmologist since many species with which diamond could be confused do contain liquid inclusions which are common in most mineral species.

Though for the sake of completeness the RI was included among the properties, the gemmological refractometer will not give a reading for diamond since the RI of the glass used in it is less than 2.0. As a member of the cubic crystal system, diamond is singly refractive: a stone offered as diamond and showing double refraction (back facets and inclusions doubled) cannot be diamond. Dispersion (the breaking up of white light into its component spectrum colours) is high in diamond, and is commonly known as 'fire': interestingly, for a species of so high an RI, the dispersion is less than might be expected and is exceeded by that of the simulants rutile and strontium titanate, discussed elsewhere.

Thus a stone with rainbow-like flashes at every turn may not be diamond: synthetic rutile or strontium titanate (the latter often forming part of a composite) has a much higher dispersion than diamond, and rutile often shows an off-white to yellowish colour which may suggest Cape diamond, although the lens will soon show the very high birefringence of rutile. For the dangerous imitation cubic zirconia, see the chapter on diamond (Chapter 7). Till then, remember that a very clean bright stone is always suspicious!

Corundum

The aluminium oxide corundum includes the varieties ruby and the different colours of sapphire. Ruby and blue sapphire also provide star stones. Corundum is second in hardness only to diamond among the gem minerals, 9 on Mohs' scale: crystals occur as hexagonal bipyramids (two hexagonal pyramids joined together at the base) or as flat (tabular) crystals with a hexagonal outline. Corundum is dense, with an SG of 3.99, so that a small stone will feel heavier than you expect (this effect is known as 'heft'): there are two RI values, as corundum, a member of the trigonal crystal system, is birefringent – they are 1.76 and 1.77 with a birefringence (found by subtracting the lower from the higher reading) of 0.009.

The coloured varieties of corundum (colourless corundum is rare in nature, and examples are most likely to be Verneuil synthetics) commonly show the characteristic absorption spectrum of the element giving them their colour. Thus ruby will show a chromium spectrum, and green sapphire the characteristic spectrum of iron, an effect shown also by blue and some yellow sapphires though often less clearly. Most ruby will fluoresce red under LWUV, but since iron 'poisons' luminescence, some specimens and most sapphires do not usually respond to UV: none the less, some iron-poor yellow stones, often from Sri Lanka, will show an apricot–yellow fluorescence.

Sadly, the many synthetic products put on to the market to imitate corundum varieties will behave in the same way as the natural material, so that the SG and RI are little help in distinguishing one specimen from another. Hardness would scarcely be used as a test anyway so, as usual, we are forced to learn about the inside of the stones.

Corundum contains a variety of mineral inclusions, and, as we say many times through the book, synthetic products show no solid material. Depending upon mode and place of formation, ruby may contain crystals of corundum itself, recognizable by their hexagonal outline, rhomb-shaped crystals of calcite or dolomite, or needle-like crystals of rutile intersecting at 60° which, if appropriately positioned, lead to the formation of the six rays of a star. Enmeshed rutile crystals are traditionally known as 'silk'. Small octahedra of spinel may be found (the two minerals share the same formation processes), and noticeably well-shaped hexagonal crystals of apatite indicate that calcium has played a part in the formation history of some corundum. Liquid inclusions often take the form of flat planes made up of shreds and tubes: the names 'fingerprints' and 'feathers' have long been used for these structures. When the liquid residues are violently twisted, it is likely that the 'liquid' is really flux residue from the flux-melt growth process by which many synthetic rubies and a few other corundum colours have been manufactured.

Rubies from Thailand and probably from other places with similar geology may contain no rutile crystals: special mention is made of this because their absence from a specimen has often provoked comment and incorrect attribution as a synthetic product. Such stones often contain characteristically shaped crystals of one of the feldspar group minerals: they consist of two individual crystals 'twinned', joined together according to one of several 'twinning rules'. Admittedly the identification of mineral inclusions is a task for the mineralogist, but even to see some solid object with a distinct shape is enough to distinguish a natural corundum of this origin from its synthetic counterpart. Thai stones may also contain long parallel 'lines', which are signs of twinning, not this time in an inclusion but in the corundum host itself.

Colour distribution in natural corundum is just as idiosyncratic as in other gem minerals. A specimen with an even coloration should be suspected as a synthetic until proved otherwise – and this is unlikely! Due to the stop-and-start manner in which natural crystals grow, an even coloration is virtually ruled out, whereas when all the parameters of crystal growth are carefully monitored and controlled, colour, as one of them, can be nearly perfect. Both natural and synthetic corundum show pleochroism (different overall colour seen with different directions of viewing), and no useful information can be provided with the dichroscope.

Perfect colour distribution is not the only feature of synthetic coloured corundum specimens: solid inclusions are absent and replaced either by curved growth bands and curved areas of colour, or by twisted veils of residual flux. Verneuil-grown stones will contain well-rounded gas bubbles with a bold outline – apart from the glasses, no natural specimen ever shows them.

Natural star corundum most frequently shows good colour and poor star (off-centre, weak or absent rays), or the other way round. In addition, weight saving is achieved by leaving the back of the stone in a rough state instead of polishing it flat. Synthetic star corundum is well coloured with a bright, central and complete star – and a flat base. There are, of course, natural star rubies and blue sapphires with superb colour and translucency, combined with a near-perfect star – but they are certainly rare. Oddly, star stones are much more favoured in some countries (Japan in particular) than in others.

While synthetic corundum echoes the colours of the natural material, one example is not found in nature. Usually spoken of as 'synthetic alexandrite corundum' or as synthetic corundum imitating alexandrite (sometimes as alexandrite), the material looks far more like amethyst than alexandrite (this seems not to have been noticed by some gemmology books). None the less, it is as alexandrite that this material causes most trouble though it is quite attractive. Natural alexandrite will be discussed in the appropriate section, but as far as its membership of the corundum family goes, this slate-blue to purple material (these are the pleochroic colours) is easily recognized by the experienced eye and, to make quite sure, the strong isolated absorption band at 475 nm (caused by vanadium, which is used as a dopant to achieve this particular colour) is diagnostic. It is amazingly persistent and one of the most successful of the synthetic stones – which is a credit to the simple Verneuil growth technique and to the marketing methods of the manufacturers. Should this material turn up for testing in a situation where only the lens is available, this variety of Verneuil synthetic shows the curved growth lines far more clearly than the others: in yellow Verneuil sapphires they are virtually invisible unless special techniques are used.

Natural amethyst, we said, could very easily be imitated by the vanadium-doped corundum, but an RI test would soon show up the difference.

Though we shall find later in the book that some natural rubies from Thailand are now heated to improve their colour, 'old timers' from this source often show a darker red, sometimes with a hint of brown, than stones from, for example, Myanmar. Such stones will show little or no fluorescence because of their iron content. In using the crossed-filter technique, in which the specimen is illuminated by monochromatic blue light and then viewed through a red filter, the iron-rich stones will glow a much more subdued red than the chromium-rich, iron-poor rubies. Some rubies from East Africa also show a subdued red, probably for the same reason.

Spinel

This gem species is not so well known as corundum, with which it is often found. Spinel is a magnesium aluminium oxide, and is a member of the cubic crystal system, in which it occurs as octahedra or as characteristic butterfly-shaped twins. Spinel is hard, over 8 on Mohs' scale, is not brittle and shows no easy cleavage. The SG of natural spinel is 3.60 and the RI is 1.718. As a cubic mineral there is no birefringence.

Red spinel is valuable, and varies in colour from a near-ruby to a most attractive orange-red. This colour is due to chromium, and the absorption spectrum is characteristic of that element, showing a group of emission (coloured) lines at the red end of the spectrum (chromium-rich ruby shows only two, and they are so close that with a small hand spectroscope they will appear as one). Red spinel can be distinguished from ruby by the absence in spinel of two absorption lines set close together in the blue part of the ruby spectrum. Red spinel glows red under LMUV.

While until recently blue natural spinel was believed to be coloured only by iron, some specimens owing their colour to cobalt have been found in Sri Lanka. Cobalt-coloured blue stones glow red when viewed through the Chelsea filter, and any blue spinel behaving in this way could safely be classed as a Verneuil synthetic: however, this is not a safe diagnosis today. Star spinel is seen from time to time but is usually pink rather than strong red, with a four-rayed star. Colours other than red, blue or pink are not usually found in natural spinel, and if a specimen proved to be spinel shows a different colour, it will be found to be a Verneuil synthetic. Colourless natural spinel is virtually unknown, so any colourless crystals or faceted stones proved to be spinel will undoubtedly be artificial. By 'proved' the gemmologist means that the RI will be that of synthetic rather than natural spinel, one case of the properties varying between the natural stone and its stone and its synthetic counterpart: this is not normally the case with synthetic gemstones.

Natural spinel shows a variety of inclusions, the chief of which is spinel itself, occurring often as chains or strings of octahedral crystals: they are often another member of the spinel mineral group, the iron spinel, magnetite. Spinel from Sri Lanka may contain zircon crystals surrounded by characteristic stress marks which are often known as haloes.

We shall see later that spinel crystals grown by the flux–melt technique take octahedral form and a variety of colours. Many if not most natural spinel crystals do not show octahedral form since they may have been abraded by the action, occurring more frequently as shapeless lumps. Again, colourless spinel crystals grown by the flux method are nearly inclusion-free, and in any case colourless natural spinel is hardly known since, as with many minerals, elements able to give colour are around as the crystal begins to grow. The colours of Verneuil-grown synthetic spinel, as we shall see, are intended to imitate other gemstones, such as aquamarine, blue zircon and peridot rather than spinel itself.

Beryl

Beryl is a beryllium aluminium silicate, crystallizing in the hexagonal crystal system and including the major gemstones emerald, aquamarine, and golden and yellow beryl (the name helidor is sometimes used for the golden stones and sometimes, in hope, for the pale yellow ones). Pink beryl is called morganite, and there is also a red beryl, so far yielding only small faceted stones. Colourless beryl is occasionally faceted for interest, as crystals can be large. The attractive colours of the beryl gemstones, especially of emerald, make synthesis a commercial proposition even though the cheap Verneuil flame-fusion technique is not suitable for the growth of silicates.

Beryl has a hardness of just over 7 on Mohs' scale, and the SG range covering all the coloured varieties is 2.66–2.80: the range for the RI is 1.56–1.60 with a birefringence of 0.004–0.008. Though beryl does not cleave easily, stones are notably brittle, and for this reason the use of ultrasonic cleaners is not recommended. Pleochroism is fairly distinct, and absorption spectra are useful in separating beryl gem varieties from other stones of similar colour: emerald shows the characteristic spectrum of chromium, and aquamarine that of iron. We should note that green beryl, not coloured by chromium, is found at a number of places: however close the resemblance, green beryl containing no chromium cannot be called emerald. Vanadium-coloured beryl can be indistinguishable from emerald but does not, of course, show a chromium absorption spectrum.

As always, familiarity with inclusions is the best way to learn about emerald and its simulants. Since synthetic emeralds are made by the flux-melt process and also hydrothermally (this process echoing one of the nature's methods of creating emerald), they contain more complicated furniture than Verneuil-grown corundum or spinel.

Colombian emeralds show characteristic crystals of brassy yellow pyrite, rhombs of calcite, occasional dark-yellow to brown parisite crystals, and the three-phase inclusions, well known to gemmology students and hopefully others, made up of angular liquid patches in which float a well-rounded gas bubble with bold outline and a cubic crystal of sodium chloride. None of these elements is found in any type of synthetic emerald. Emeralds from other locations are equally rich in solid inclusions: bamboo-like actinolite rods characterize stones from the Urals, and short, slender tremolite crystals are found in stones from Zimbabwe.

Emeralds from Pakistan and Brazil more commonly contain liquid inclusions, and these can be quite easily mistaken for twisted veils of undigested flux, as shown by most synthetic emeralds – or, of course, the other way round. More details of the inclusions of beryl can be found in monographs on the gem materials, but here we need only say that aquamarine is far less included than emerald: this is partly due to aquamarine forming quite large crystals, so that the lapidary can facet stones from areas of the crystal with few or no inclusions. When aquamarine is included, long thin hollows (sometimes known as 'rain') are seen parallel to the long direction of the original crystal.

Generally, as we shall see in the chapter on emerald synthesis and imitation (Chapter 9), there is no real difficulty in distinguishing synthetic emeralds from their natural counterparts when familiarity with inclusions has been developed. There are simpler tests with which the gemmologist can begin – though they are not diagnostic. One is that, since crystal growers can control the chemistry of their crystals, iron-free and chromium-rich stones can be grown. Such stones when viewed through the Chelsea filter will glow a very bright red (the filter was developed to counter the threat of the imitation, then the synthetic emerald). In general, too, most synthetic emeralds have a lower SG and RI than natural ones.

Quartz

Quartz, with the simple composition silicon dioxide, is routinely grown in amethyst and citrine colours, as colourless rock crystal and is occasionally treated to give transparent pale green or blue stones. All synthetic quartz is grown by the hydrothermal method, described elsewhere. In nature, the simple unvarying chemical composition of quartz means that there is no range for major constants: the SG is 2.651 and the RI is 1.544–1.553 with a birefringence of 0.009. The hardness of quartz is 7 on Mohs' scale.

A member of the trigonal crystal system, quartz forms prismatic crystals with pointed rhombohedral terminations (they resemble pyramids): twinning is very common, and is either visible or invisible. While an excellent test for transparent quartz involves the polariscope, between the crossed polars of which quartz will show a unique and diagnostic interference figure, this is the same whether the specimen is natural or synthetic. It is in fact quite difficult, in the absence of clear natural inclusions, to decide on the origin of amethyst, as we shall see later.

While synthetic quartz is more or less inclusion-free, natural specimens may contain yellow–brown rutile needles, green crystals of tourmaline, blood-red hematite platelets, greenish chlorite or yellow-to-orange crystals of goethite. Star quartz is known, occurring in the rock crystal and rose quartz varieties, and profuse green crystallites of fuchsite (one of the mica group minerals) give green aventurine quartz. None of these substances is found in synthetic quartz.

As far as the crypto-crystalline quartz varieties are concerned (these are made up of extremely small crystals) they are not manufactured, and their only danger to the gemmologist is when they are artificially coloured: even then, no great financial loss should be involved.

Opal

While synthetic opal has been on the market for some years, an equal danger to those gemmologists and jewellers who do not habitually handle opal comes from the many opal imitations, some of which are ingenious and deceptive. They will be examined, together with synthetic opal, in the opal chapter (Chapter 11).

Opal is silica with a variable amount of water: it is soft, usually about 6.6 on Mohs' scale, non-crystalline and with an SG around 2.10. While the RI is in the 1.45 area, it should not need to be established since identification can be made by other means: in any case, the use of a liquid on a porous material may alter its appearance.

Opal with a play of colour may have a dark background (black opal) or a light one (white opal). Both these varieties are translucent to a strong light. Fire opal may have a play of colour, but stones are transparent with a red–orange–yellow colour. Water opal is transparent and colourless with the play of colour suspended within: this is an especially beautiful gemstone.

Opal is one of the gemstones that may be even more valuable in the rough state than when it is fashioned. For this reason, gemmologists need to become familiar with the rough material, which sometimes occurs in interesting and attractive forms: they may include opalized plant or fossil material, as well as the nut-like specimens with a kernel of opal. Boulder opal is brown-banded ironstone (a sandstone) with some included opal.

A look through the somewhat pedantic old-style gemmology textbooks will soon show that the number of names used for opal varieties is very large. They can be ignored in the task of deciding whether a particular specimen is opal or an imitation.

Alexandrite

Alexandrite is one of the gem varieties of the mineral chrysoberyl and, with cat's-eye, is one of the more expensive gemstones. Fine specimens of alexandrite appear raspberry-red in incandescent light (bulbs or candle-light), and green in daylight or under white strip (fluorescent) light. This colour change (nothing to do with pleochroism, though two of the three possible pleochroic colours are similar) has always attracted connoisseurs of gemstones, and it is not surprising that alexandrite has long been imitated and, more recently, synthesized. The cat's-eye chrysoberyl has not yet been synthesized as it is fairly easy to make a convincing imitation.

Chrysoberyl is a member of the orthorhombic crystal system and is hard and tough, 8.5 on Mohs' scale. The SG is 3.71–3.72, and the RI usually near 1.74–1.75 with a birefringence of 0.008–0.010. There is no cleavage, and faceted stones, which include a beautiful yellow-green variety, are particularly bright, though this effect is not so apparent in the naturally darker alexandrite. Chemically, chrysoberyl is beryllium aluminium oxide.

Testing by normal gemmological methods is straightforward, but if there are only magnification facilities available the gemmologist should look for signs of flux growth in the synthetic alexandrite (which will have the same SG and RI as the natural material in any case) or for inclusions characteristic of natural chrysoberyl. These are not always easy to see: specimens may show liquid-filled cavities or short needles; step-like effects, due to twinning, are sometimes visible. The synthetic and imitation materials will be discussed later.

Topaz

Even today some dealers do not clearly distinguish between topaz and the similarly coloured citrine variety of quartz. But to us (and to the gemmologist) topaz is aluminium fluosilicate rather than silicon dioxide: topaz is harder than quartz at 8 on Mohs' scale (quartz is 7), and has a higher SG (3.53 compared to 2.65) and RI (1.62–1.64 against 1.544–1.553). So the simple methods of gem testing can quite easily distinguish between topaz and quartz and between topaz and other gem species.

Topaz gemstones are not only found in the orange–yellow–brown colours with which the name is usually associated but also in pink and blue. The pink stones occur naturally, and I well remember finding superb dark purplish-pink crystals in the North-West Frontier Province of Pakistan: the colour is stable. Other pink crystals are found in Brazil, but some brownish material can be heated to give a stable pink. Blue crystals are mined in Brazil and Zimbabwe, but many blue stones now on the market are a much darker blue than natural topaz and have gained their colour from treatment.

Topaz cleaves very easily, and the presence of rainbow-like markings inside a stone is a danger sign as well as suggesting that a specimen is topaz. The colours are hard to see in the darker material.

Chapter 7

Diamond

As we read about diamond we shall encounter *diamond types*. These are conventionally written type I, type IIA and so on. There is no need for the reader to investigate types in any detail, and a full summary of today's classification can be found on p. 52. Nonetheless, it is impossible not to mention types as we go along, but today's *fine divisions* (which could change at any time) are not *always* given, because they are sometimes irrelevant to the investigation reported. So when you read 'Type IaB stones are . . .' and then, later, 'Type Ib stones are . . .' (thus encountering the special before the basic class) you are not being denied fuller information!

Since diamond is the most important if not always the most expensive gemstone, we include in our discussions many major properties common to gemstones in general and the ways in which they can be identified. In particular, we introduce the topics of the amorphous and crystalline states, upon which so many properties depend. We also describe, beginning with the most simple examples and progressing to gemmological and laboratory testing, the ways in which the amateur or busy jeweller, the gemmologist and the museum scientist set about identifying and distinguishing one gemstone from another. Full details of the tests and instruments used are given in Chapter 5.

All the synthetic, imitation and enhanced gemstones will be covered following the same pattern, so that readers wanting the simplest tests should look for the sections in which they are described and not worry too much when the text becomes more complicated.

Non-crystalline materials imitating diamond

Glass: the commonest imitation of diamond

If you are going to imitate a gemstone by a man-made product, by another natural material or by altering the appearance of some other stone, you might as well start with the imitation of diamond by *glass*. Most gem-quality diamonds are colourless so the production of your substitute stone does not have to involve ways and means of inducing diamond-like colour: as long as your product gives something like the 'fire' of diamond (the flashes of spectrum colour from the small upper facets), glass will do. Everyone knows about diamond so you do not have to establish an entirely new name – most of them sound cheap anyway, and the buying public is conservative. So why not stick to glass? Easy to make and to colour if you want to, pleasing appearance, and easy to match with other glass specimens in size and colour. Sadly, though, glass is soft and fractures easily; it is also light, and a necklace may not hang well. Moreover, glass is quite easy to spot once suspicions are aroused.

None the less, glass is versatile. Not only can you make whole stones from it, obtaining a faceted effect either from a mould or, in the better qualities, faceting by grinding like a 'real' stone, but glass can also be used in composites, forming either the top (crown) or the base (pavilion) and sometimes lurking elsewhere, often filling fractures. In what simple ways can you distinguish diamond from a glass?

Outside the specimen

The facets of diamond are uniquely sharp if properly polished. Using the simple standard 10× lens you will see that facets on glass imitations are never as sharp, and the cheaper moulded glasses show rounded facet edges. Larger diamonds may show facets additional to those of the standard brilliant: they will have been added either to improve brilliance or to make up for errors in polishing. With glass you would not expect extra facets since it would not be worth the cost of polishing them. Still, with the 10× lens a faceted diamond often shows a girdle left unpolished and small triangular markings (trigons) against the frosted unpolished surface. Glass will not show the frosted girdle nor the trigons, but many diamonds have polished girdles which will show no trigons. Glass is brittle, and small (or large) fractures abound on faceted edges and in the girdle area. These fractures are known as conchoidal from their shell-like appearance, and while most other gemstones also fracture in this way, glass shows far more of these characteristic signs of breakage.

Glass is a poor conductor of heat since it does not have a regular internal atomic structure. Diamond, on the other hand, has one of the most compact and interlinked atomic structures known and is the best conductor of heat known in nature. For this reason a very simple test can in theory be carried out to distinguish diamond from glass. Touched with the tip of the tongue diamond feels much colder than glass: however, since your specimen may previously have been tested with chemicals, this test is not recommended! In any case, it is not very precise.

Because diamond has a particularly compact and interlinked atomic structure it also possesses a particularly smooth surface when polished. A drop of water placed on the surface will not spread as it does with glass and other diamond simulants. In the same way the 'bead test' allows a line drawn across the diamond surface with a special pen to remain homogeneous and not break up into tiny beads.

Inside the specimen

If with the lens you are able to look into your specimen (let the lamp shine on the stone, not into your eyes), most specimens of glass will show characteristic swirl marks, denoting incomplete mixing of the components, and gas bubbles. These, too, arise from the manufacturing process, and are especially notable for their well-rounded (sometimes flattened) shape and thick outlines. Gemmology students on first looking into a stone often designate any small isolated object as a 'bubble': they are much more likely to be tiny crystallites, as single bubbles are rare in natural materials. Glass will not show the natural solid inclusions seen in so many diamonds, and this is usually the best way in which to distinguish glass (and other man-made substances) from any natural mineral. Beware, though, of glass whose components have begun to crystallize (glass is often called a 'super-cooled liquid'): the tiny crystallites could easily be mistaken for inclusions in a natural material.

Glass in which recrystallization has begun is known as 'devitrified glass'.

No man-made material will show natural solid inclusions. When you look into a diamond you will almost always find that the stone is not 'flawless', in the sense of included material being absent, but that it will contain a small amount of solid mineral matter. It takes a skilled gemmologist to find out exactly what these inclusions are, but their mere presence rules out glass. Among the natural mineral inclusions found in diamond is diamond itself, whose unique 'adamantine' lustre rivals that on the surface of the polished stone: diamond inclusions may even show something of the mineral's characteristic crystal shape, with parts of the double pyramid (octahedron) visible, though this would be rare. Quite often diamond inclusions in diamond are quite large and of course detract from the appearance of the faceted stone. Among other mineral inclusions are pale green olivine, green diopside and red garnet. These tiny mineral specimens are always very small and hard to see at low magnification. Liquid inclusions, which we shall meet in other gem species, are notably absent from diamond, whose formation takes place under conditions of such great heat and pressure that no liquid phases can survive. No other colourless stone, much less glass, contains these particular mineral inclusions.

While diamond crystals do not break easily (in the sense of fracture rather than shattering) and of course can be scratched by no other substance, they do cleave: cleavage is breaking along directions specific to the crystal system of a particular mineral. These directions are lines of atomic weakness, and when cleavage does take place (it is used by the diamond polisher to eliminate unwanted inclusions lying in an appropriate part of the crystal) it leaves smooth surfaces behind. If cleavage has begun inside a stone, rainbow-like markings will be seen under ordinary magnification.

The tests described above can be carried out whether or not the diamond is mounted. Small stones set in a complicated piece of jewellery may include both diamonds and glass, and not all jewellers or gemmologists have time to test all of them when there may be a hundred or more. We have to remember, too, that open settings allow dirt to accumulate on the backs of stones, and when viewed through the crown the dirt particles may resemble natural inclusions. Dirt may accumulate in smaller amounts behind the stone in a closed setting, and less profuse dirt may suggest a mineral inclusion more subtly.

Coloured diamonds

Coloured diamonds are hard to test because the nature of their colour is often hard to determine, but we shall look at this complicated problem later. Distinguishing coloured diamonds from glass imitations using simple tests is not very different from testing colourless specimens: diamonds have quite a restricted range of colours while glass can take on whatever colour the manufacturer decides. Diamonds may be yellow, green, blue, brown, pink to very nearly red as well as colours rather more difficult to define such as 'champagne' or 'chartreuse'. Interestingly, when diamonds are treated to alter or improve their colour, the final colour is rarely if ever one not seen in natural diamonds: a pity, since this would save a great deal of testing! Some diamonds which are neither white nor a deep yellow have traditionally been called 'Cape' stones, and while these are important in the colour alteration context, they can easily be mistaken for off-white glass, and, of course, vice versa.

When diamonds have had their colour changed by artificial means the 10× lens can identify them for you in one context only. We shall meet this again below, but stones with an umbrella-like marking showing above the culet (the point at the base of the stone or the area where such a point would be if there were to be one) will have their colour (usually green or blue) from treatment in a cyclotron.

Testing with easily-operated instruments

The unique structure of diamond enables the polisher to obtain a surface finish of near-optical perfection. Diamond thus has an unrivalled power to reflect light, and a simple device, using a beam of infra-red radiation, quickly and easily compares the surface of an unknown specimen to the programmed-in diamond surface reading. This enables the operator to tell instantly whether the specimen is a diamond, a piece of glass or some other stone. All the 'non-diamonds' are lumped together in the simplest of these instruments so your reading (or audible signal) will tell you if your specimen is diamond but will not indicate the true identity of an imitation.

We have seen that diamond by virtue of its atomic structure is a better conductor of heat than any other substance. While the 'tongue test' might be found useful it is hardly precise, and it is not surprising that a simple instrument has been devised to compare the heat conductivity of diamond with that of diamond imitations. This is the thermal conductivity tester, consisting of a heated copper probe which, when applied to the surface, measures the speed at which heat is conducted away by the specimen. Since diamond completely out-performs all other stones there is no need to say what an imitation is: all the thermal conductivity tester does is to tell us 'diamond/not diamond', as with the simpler reflectivity meters.

Both these tests are easy to perform, and while there are some possible pitfalls, which we shall look at later, the instruments are cheap and quickly rule out glass. With other diamond imitations a little more work may be needed.

Summary

So far we have seen how the unique crystal structure of diamond separates it in hardness, power of surface reflection, unique thermal conductivity and interior furniture (inclusions) from the simplest and commonest of its imitations, *glass*. On the way we have met several instruments whose use will become familiar, and we have made intriguing references to other materials and tests.

Readers are not yet equipped to distinguish diamond from all its imitations, nor from synthetic gem-quality diamond, now a feature of the markets, nor from diamonds whose colour has been altered or improved. For all these materials, gemmological tests are needed (sometimes even more advanced techniques have to be called in). Most of the gemmological and some of the advanced tests are briefly described elsewhere in the book.

Crystalline materials imitating diamond

We looked at glass first because it is by far the commonest imitation of diamond. Glass has another property not shared by many other gem materials: it is not *crystalline* (only opal among all other known materials shares this property). This non-crystallinity is known as the *amorphous state*, and it is surprising that of the very few known amorphous materials, two of them should have ornamental applications. The lack of a regular internal atomic structure in amorphous substances leads to their poor heat conductivity and to other detectable properties, and is useful to gemmologists since it can be detected by gemmological instruments.

So we know that, apart from glass and opal (an unlikely substitute for diamond), all possible simulants, in common with nearly all known organic and inorganic substances known on Earth, are crystalline.

Crystals have a regular internal atomic structure and so can possess a variety of properties: these include change of colour with direction of viewing (*pleochroism*), directional hardness, and the property of *birefringence*, in which a ray of light on entering the material, is split into two rays each traversing it at different velocities. Birefringence does not occur in all crystals as we have already seen, and as it so happens diamond is one of the non-birefringent substances. Non-birefringent materials are called *isotropic* since rays of light entering the crystal are not split into two but traverse the crystal at the same velocity irrespective of direction. Birefringent stones are *anisotropic*.

Gemstones are always keen to catch you out: glass is another isotropic substance, so here at least diamond and glass share a property. The most successful diamond simulant, cubic zirconia (CZ), is also isotropic, as is yttrium aluminium garnet (YAG), another simulant, discovered prior to CZ. Both CZ and YAG are artificial substances with no natural counterparts so they cannot strictly be called synthetic.

Testing with easily operated instruments

With the 10× lens *anisotropic* stones, when their birefringence is strong, will show the back facets doubled when they are viewed through the table (top) of the stone. If this effect is observed (the facet edges look like tramlines, the rails far apart if the birefringence is strong and closer together if it is small), the stone cannot be diamond because diamond is *isotropic*. The 10× lens will show back facet doubling in the natural mineral zircon (no relation to CZ), which is otherwise a promising simulant for diamond. The same effect will be seen in the far less common *scheelite* (calcium tungstate), which, since the mineral is an important ore of tungsten, does get faceted now and then. Like zircon and diamond, scheelite has a high dispersion (the power of breaking up white light into its component spectrum colours), so at first sight you could be deceived by either material.

Other, commoner, colourless stones sometimes offered as diamond while having no particular resemblance to it include *synthetic corundum* (small birefringence) and *synthetic spinel* (isotropic). We shall meet them in other contexts where they play a much more important role in imitation.

So far the crystalline diamond imitations we have noted can be assessed with reasonable confidence using the 10× lens – so long as birefringence is present. What happens with isotropic materials such as CZ and YAG? Here the lens will tell you little about the optics of a specimen, but, as we found above, diamond usually contains inclusions of itself or of other minerals: neither CZ nor YAG, manufactured in the laboratory or factory, can show any natural inclusions, and in both cases reflectivity from their surfaces is much less than that from the surface of diamond and the reflectivity meter will make the distinction clearly and easily. *Synthetic spinel* is another matter, since specimens are often too small to be tested by the reflectivity meter and, like other simulants of diamond, it may occur as tiny stones in a complicated piece of jewellery. Here the thermal conductivity tester with its delicate copper probe is the best instrument to use.

Testing with gemmological instruments

The refractometer presents to the observer a reading which expresses the ratio of the velocity of light inside the stone to the velocity of light in air. Inside the stone the velocity of light is lower since atoms of the constituent elements in the stone get in the way and retard it: how this property is utilized has been explained in detail in the instruments chapter (Chapter 5). Since each gemstone has a different chemical composition and thus a different atomic structure, the reading, known as the *refractive index* (RI), will also be different, and the refractometer is a quick and useful way of distinguishing one gemstone from another – providing certain conditions are fulfilled. Not all gemstones give a reading since operating conditions in the refractometer limits the range of stones that can be tested. Sadly, diamond is one of them! Diamond has an RI of 2.417 and cannot be tested on the standard refractometer, whose upper limit is approximately 1.77. However, many of its imitations will give a reading and here are some of them:

Synthetic colourless corundum: 1.76–1.77
Synthetic colourless spinel: 1.728
Glass (typical range): 1.50–1.70.

You will see that while corundum has two readings, synthetic spinel has only one: this is because corundum is *anisotropic* and spinel *isotropic* (see above). For interest, the more dangerous imitation CZ has an RI of 2.17, and its predecessor YAG a value of 1.83, these of course falling outside the refractometer range and needing other tests to establish their identity.

A less common though no less dangerous substitute is isotropic *strontium titanate* with an RI of 2.41, very close to that of diamond. Its power of *dispersion* (of breaking up white light into its spectrum colours when the crystal is faceted) is higher than that of diamond, but the material is far softer and can be marked by the point of a needle, easily seen using the 10× lens. When strontium titanate is faceted and used as the base (pavilion) of a stone with a harder, colourless, transparent top (crown) the resulting *doublet* is very hard to spot, especially if the stone is small. While the 10× lens can show the joint between the two portions, you have to be *ready for composites*!

While strontium titanate is sometimes doped (has other elements added to its normal composition) to give colour, there is another diamond imitation, again with a dispersion higher than that of diamond, which is never truly colourless but has a yellow cast. You might very well think 'but then it might look like a Cape (off-white to very pale yellowish) diamond!' Fortunately, this very common imitation, *synthetic rutile*, is *highly birefringent* and thus shows the back facets clearly doubled when viewed through the table of the stone. Like strontium titanate, synthetic rutile has too high an RI for the gemmological refractometer to measure. But since its 'normal' off-white colour is not much use for ornament, despite the high dispersion, dopants ('foreign' chemical elements) are added to give, blue, yellow or brown colours which are distinctive and which could be taken as diamonds of similar colour. So yet again the 10× lens plays a vital role in showing the doubling of the back facets.

The 10× lens will also identify the rare *diamond doublet*. This is two pieces of diamond joined together as an apparently single faceted stone. The aim is not only to deceive but to use pieces of diamond too small in themselves to make a saleable stone. Looking through the table facet, the table itself can be seen reflected lower down inside the stone, and a small object placed on the table will also be seen on the false table inside. The diamond doublet is rare but examples do exist!

While isotropic CZ and YAG cannot be tested on the refractometer and show little under magnification, both species and the YAG analogue gadolinum gallium garnet (GGG) pose quite serious threats to the gemmologist. All have been doped with different elements to give a range of attractive colours, though colourless material is common (GGG, more expensive to make, less so). We shall look at the nature and production of these materials elsewhere, but considering them as imitations of diamond we need to use the reflectivity meter to diagnose them satisfactorily. The meter normally shows either 'diamond' or 'not-diamond' (though one long-established model does identify the non-diamonds, most users do not want to know what a stone is, if it is not diamond).

We are much better off when these products occur with dopants giving colour. Superficially, the colours are much brighter than the colours usually seen in diamond, but jewellers would be rash to judge stones on colour alone.

Gemmologists use a simple *spectroscope* which is especially useful when the specimen is coloured. We have looked more closely at its construction and operation elsewhere (see Chapter 5) but, described once more in simple terms, the instrument presents to the observer a ribbon of spectrum colours, conventionally in Europe positioned with red on the left and violet on the right (the reverse position is used in the USA). When certain elements are present in a specimen which is examined in strong transmitted or reflected light, dark vertical lines or bands are seen superimposed upon the ribbon of spectrum colours. The number, position and thickness of these *absorption* bands or lines vary from species to species, thus giving a simply obtained picture of what elements are present. When you know this you have a very good idea of what the stone is.

The Cape (off-white to pale yellowish) diamond shows one such dark line in the blue portion of the spectrum. This line, at 415.5 nm, is diagnostic – that is, the presence of the line proves that the stone is a Cape variety of diamond, and no other test is needed. Looking ahead a little to the topic of the alteration of the colour of diamond, the presence of this absorption line for a fine yellow stone would show that it began life as a poor-relation Cape stone, and has therefore been treated.

Absorption lines shown in the spectrum of a coloured gemstone offered as a diamond will often be very fine and may extend throughout the spectrum. Such an effect is seen in no coloured diamond and would undoubtedly indicate an artificial product. The effect would also be rare in any other coloured gem species, so the spectroscope is a very useful and easily operated instrument for those engaged in all types of gem testing. These fine-line absorption spectra are produced by the crystal grower adding *rare earths* to obtain particular colours. While we have already said that the colours produced are generally much brighter than those seen in coloured diamonds, not everyone will know this, so a coloured CZ, YAG or GGG simulant should be in the back of the mind each time you are offered 'unusual coloured diamond'.

While not all gemmologists have access to sources of ultra-violet radiation (long-wave and short-wave ultra-violet, LWUV and SWUV) many collectors of minerals find UV indispensible for their specimen testing. We have seen how they work, but now we need to know what use they are in identifying diamond and its dangerous imitators.

By no means all diamonds fluoresce, but, in general, Cape diamonds (off-white to yellowish diamonds showing the diagnostic absorption line mentioned above) show a blue fluorescence under LWUV and, after the UV is switched off, a yellow phosphorescence persists for a few seconds. This effect is diagnostic for diamond, and it is a pity that only Cape stones can be tested in this way! While some diamonds show a greenish-yellow fluorescence with yellow phosphorescence, these effects are not so common and are less easy to observe and evaluate. Apart from the Cape stones, diamonds show a variety of fluorescent effects. Since colour-enhanced diamonds show some of them, they will be discussed under that heading (see p. 47).

Synthetic colourless spinel is dangerous when used in small sizes as diamond replacements – for example in place of a missing diamond in an eternity ring. While textbooks confirm that its dispersion is much less than that of diamond, when in place this difference is less easily perceived. If the ring is placed under SWUV any synthetic spinel will give a sky-blue fluorescence while diamond will not respond. However, some synthetic gem-quality diamonds may respond in a similar way, so the test needs careful backing-up if anything out of the way is suspected. Suspicion is the best gemmological instrument, and today all gemmologists and jewellers need to suspect every stone: it is what scientists have always done. A piece of jewellery set with diamonds throughout and with *each* stone giving a uniform fluorescent response to UV or to X-rays should be regarded with maximum suspicion.

We must not expect all the stones we encounter to be mounted or even faceted. Jewellers all come across the occasional collector who comes in with diamond crystals! This is rare but does happen. A much more likely subject for testing would be a loose colourless diamond, either purchased in that state or removed for some reason from a piece of jewellery. Having exhausted the 10× lens tests described above and finding no help from UV or from the spectroscope, it might be worth testing the specimen for its *specific gravity* (SG). While gemmology textbooks pay great attention to this essentially simple test, there are many snags about its routine use, one of the worst being its slowness and messiness. We have met the test in the chapter on gemmological instruments (Chapter 5), but here we will just repeat that the SG of a body is the ratio of its weight to the weight of an equal amount of pure water at 4°C (at which temperature water is at its densest). Whatever version of the test you use, you will find that each gem species has its own distinctive SG. The SG of diamond is 3.52 while those of the dangerous imitators CZ and YAG are 5.6–5.9 (varying with composition) and approximately 4.57, respectively. In passing we may notice that CZ is 1.7 times denser than diamond: readers of the excellent Dick Francis novel *Straight* (Michael Joseph, 1989), a book with lots of gemmology in it and all accurate, will notice the formula CZ=C × 1.7. C is carbon and therefore, in this context, diamond). If you are asked for a good imitation of the same size as a 1 ct diamond, you would have to provide a 1.45 ct strontium titanate, a 1.30 ct YAG or a 1.70 ct CZ.

SG figures that may be found useful are synthetic spinel 3.64, synthetic corundum 3.99, synthetic rutile 4.25 and strontium titanate 5.13. Glass, whose composition varies from specimen to specimen, is usually in the range 2.4–5.1, the figure depending on the elements added to give colour.

Some of the liquids used for one of the SG-determining tests may also be useful for immersing gemstones. When a gemstone is completely immersed in a colourless transparent liquid, its outline may appear bold and its facet edges faint – or vice versa, depending upon the respective RI values of the liquid and specimen. The effect can easily be seen when ice is placed in water (or gin). The RI values of both are so close that the ice is hard to see. The liquid di-iodomethane, whose RI is 1.74, is used as a major component in the liquid used to determine the RI on the refractometer, and is thus generally available in the gemmological world: when diamond (RI 2.417) is immersed in it the outline will appear very bold and the facet edges faint; a similar effect, but less marked, will be shown by synthetic spinel (RI 1.728) and by most glass (RI usually in the range 1.50–1.70). CZ and YAG, however, will show effects much closer to that shown by diamond (respective RI values are 2.17 and 1.83). If a record is needed, the stones and liquid, in a flat-bottomed glass dish, may be photographed to give an effect known as immersion contrast. In this case the effects will be reversed in the negative but the right way round in a print.

This test can be carried further: where the RI of the stone is higher than that of the liquid, a dark-rimmed shadow will be seen on a white background upon which the experiment is projected: the facet edges will appear white. When the RI of the stone is lower than that of the liquid the outline will show bright and the facet edges dark. If a colour fringe is seen, then the match of RI values is close.

If you have a microscope it may be possible to immerse diamond in a transparent glass cell, taking note of the RI of the immersion liquid and watching what happens when you raise the focus of the microscope. If the diamond has the higher index a bright line of light will pass from the liquid into the stone. In practice you will not easily find a liquid with an RI *higher* than that of diamond, so the test is only really useful in laboratories where appropriate liquids may be available. None the less, two specimens, one diamond and the other glass, may be distinguished with a liquid whose RI is lower than that of diamond but higher than that of glass. Di-iodomethane, the all-purpose liquid (RI 1.74), is useful here.

The dispersion ('fire') shown by diamond is high enough to produce flashes of spectrum colour from the small upper facets of a brilliant-cut stone. The effect is seen best in colourless stones but can be detected in coloured ones too. Unfortunately, glass can give a high dispersion! Many people believe that dispersion is a useful test to distinguish diamond from its simulants and that it is a simple test achievable by anyone. It certainly is easy to see dispersion but since the best imitations of diamond also have lots of fire (or why produce them?) they are hard to separate from diamond. Oddly enough, diamond does not show the amount of dispersion that would be expected from a material with so high an RI, and several *coloured* stones have higher values – dangerous! At this point, though there are ways to measure dispersion, they are quite unnecessary for the gemmologist and certainly for the jeweller. If you remember that lots of fire does not necessarily mean diamond, it is enough.

Summary

We compared the crystalline and amorphous states and their respective properties. Crystals may be isotropic or anisotropic according to their atomic structure, and anisotropic ones show effects of birefringence and pleochroism which form the basis of useful tests. Crystals may have directional hardness, and this is vital for diamond or polishing could never take place because only diamond can cut diamond! We looked at distinguishing diamond from its simulants by using simple gemmological instruments including the 10× lens, which will detect birefringence which diamond could never show, and the use of the refractometer. We looked at some serious simulants and composites in detail and the doping of some simulants to give them colour. Coloured stones may give a recognizable absorption spectrum which can be observed with the spectroscope: Cape diamonds give a very important absorption band which sometimes persists after treatment to alter the colour of the stone.

We saw that UV radiation can detect some diamonds and that the somewhat slow but useful SG test can also help in some circumstances. More useful is the immersion of specimens in liquids of known RI. Photography can help, and will provide a permanent record of some tests. Dispersion ('fire') is an important property of diamonds but it cannot easily be measured and when it is observed, does not always signify diamond, as some coloured stones of high RI have a high dispersion.

Synthetic gem-quality diamond

In 1970 the General Electric (GE) Company made some gem-quality diamonds, at that time as a spin-off of research into high-pressure and high-temperature materials manufacture. Some of the crystals were polished and from then the possibility of such stones entering the trade one day became a probability.

The GE diamonds, three faceted stones and five unpolished crystals, were examined by the Gemmological Institute of America (GIA), and reported in the fall 1984 issue of *Gems*

& Gemology. The time taken to grow a crystal large enough for a polished stone of about 0.5 ct was reported to be approximately one week at least.

The faceted stones were cut as brilliants: one was near-colourless, one bright yellow and one greyish blue. All the GE diamonds were inert to LWUV, but under SWUV they showed different responses. The near-colourless stone fluoresced a very strong yellow and had a similarly coloured very strong and persistent phosphorescence. The yellow stone (and the yellow crystals) were inert to SWUV while the greyish-blue diamond gave a very strong fluorescence and phosphorescence of a slightly greenish yellow. In the blue and near-colourless stones a cross-shaped pattern could be seen under SWUV.

Since among natural diamonds only the blue (type IIb) stones show a phosphorescence after LWUV and SWUV irradiation, the response of the GE stones provides a good indication of their unusual origin.

Using the direct vision (hand) spectroscope none of the diamonds showed any absorption bands, even when the stones were cooled. Many yellow diamonds begin life as the off-white to pale-yellow Cape stones whose diagnostic and persistent absorption band at 415.5 nm is so convenient a clue to origin. While some yellow diamonds will only show this band after cooling, the absence of it in the GE stones gave a further clue to something unusual, though not all yellow stones start as Cape diamonds and show the tell-tale band. Even when the GE stones were tested using cryogenic (low-temperature) spectrometry techniques the lack of absorption bands was confirmed. Examination of one small yellow GE crystal did show a band at 415.5 nm, and this may have been due to a small natural diamond used as a seed.

Some natural diamonds will conduct electricity: these are classed as type IIb. Natural type IIb stones are colourless unless some boron is present. Of the GE stones, only the blue and near-colourless specimens were found to be electrically conductive: aluminium found during examination of the inclusions may be responsible for the conductivity of the colourless specimen. At the time of the GIA report in 1984 no natural near-colourless diamonds had been reported as electro-conductive.

While SG testing by hydrostatic weighing gave an average reading of 3.51, stones containing prominent inclusions of metallic iron sank rapidly in a Clerici solution and distilled water liquid.

Under the microscope (Mark V Gemolite, used by many gemmologists) the stones all showed opaque black inclusions of the flux used in their manufacture. Shaped as rods or plates, the inclusions provide an easy means of distinction from natural diamond, which contains nothing similar. Also visible inside the diamonds were clouds of minute pinpoints of unknown composition. Growth zoning was observed in the yellow faceted stone, and similar features can be seen in the blue stones grown by GE. The yellow diamond, alone among the faceted stones, showed a triangular mark ('natural') on the girdle.

Between crossed polars on the polariscope no signs of internal strain could be seen: some natural diamonds show this too. Around the flux inclusions some signs of greyish strain haloes were visible. On the whole, the absence of signs of internal strain goes some way towards suggesting a synthetic origin.

Since the metallic inclusions were mostly iron, the GIA tried the stones with a pocket magnet. Both the blue faceted stone and the near-colourless stone showed strong attraction, the yellow stone less so. A test using a superconducting magnetometer showed that all the GE diamonds could be separated from natural stones on the basis of their magnetic properties.

As a summary, these early GE diamonds could be distinguished from natural diamonds by their magnetic properties and by their absence of strain in polarized light. For a natural near-colourless stone of the shade examined to conduct electricity would be very rare, and a synthetic origin would be strongly suggested. For natural diamonds (except for type IIb with a suggestion of a blue or grey colour) to react so strongly to SWUV with no reaction to LWUV is so far not recorded, so that a near-colourless stone without a blue or grey tint, fluorescing and phosphorescing strongly under SWUV but remaining inert to LWUV, will be synthetic.

Fancy yellow natural diamonds with no absorption bands visible with the spectroscope usually fluoresce and phosphoresce under UV. Those which do not fluoresce usually show a strong Cape absorption band: a yellow stone showing neither of these effects is probably synthetic, and a near-colourless stone with no blue, brown or grey tint, without a Cape absorption band, is also probably synthetic.

A strong fluorescence under X-rays followed by persistent strong yellow phosphorescence also suggests a synthetic origin since most natural stones give a blue fluorescence under X-rays.

The GE diamonds were the first examples of synthetic gem-quality stones to be fully reported, and the diamond trade did not seem unduly troubled by the still distant prospect of synthetic intruders into their markets. Two years after the report on the GE stones, a paper in *Gems & Gemology* for winter 1986 gave the first account of the Sumitomo gem-quality yellow diamond. Using equipment with a larger capacity than the GE workers, Sumitomo Electric Industries managed to grow many crystals up to 2 ct in size simultaneously. So great a stride forward meant that many more diamond crystals could be produced and that entry into the diamond markets on a fairly large scale was possible.

In 1985 the Japanese firm Sumitomo Electric Industries reported that they had produced diamond single crystals of gem quality on a large scale. They reported that the crystals were yellow and weighed up to 2 ct (though the first productions were smaller). The firm claimed that the crystals were very suitable for industrial use since they possessed high thermal conductivity, a high fracture strength and a relatively small inclusion content. The crystals were grown to provide materials for precision cutting tools and for heat sinks, drawing on the unique properties of diamond. Sumitomo reported at the time that they were able to produce the crystals in numbers sufficient to satisfy the demand for these applications, so that, compared with the GE experimental stones, there was a capacity to produce gem-quality crystals routinely. However, at the time of the GIA report in 1986, Sumitomo had denied any intention of expanding the sale of gem-quality diamond crystals, and by 1996 this stated intention appears to have been kept.

The crystals are said by Sumitomo to be grown at high temperatures and pressures using a flux method and a metal alloy solvent. The yellow crystals are transparent and virtually inclusion-free: growth is initiated on small seed crystals, and a controlled amount of nitrogen is included, from 30 to 60 parts per million. The nitrogen is responsible for the yellow colour. At the time of the report only deep-yellow colours had been seen although the firm claimed to be able to produce near-colourless crystals as well. Sumitomo marketed the sawn and laser-cut crystals in sizes from approximately 0.10 to 0.40 ct at (1986) prices of US $60–145 per rectangular piece. The crystals were partly polished. While uncut crystals were not to be sold, some did appear and are in a few major gemstone collections. One brilliant-cut stone of about 0.8 ct and said to have been cut from a 1.7 ct crystal, showed a cloud under the table, a step-like fracture under the girdle and a rod-shaped metallic inclusion near the culet.

Before looking more closely at the Sumitomo product (which may appear in gem-quality faceted stones at any time) it is worth looking at the classification of diamond into types on basis of the presence or absence of nitrogen. The basic classification below is capable of further refinement (p. 52), but since gemmologists will not really need details they are not given here. We should note that in the classification, *natural* diamonds are being discussed.

A very high proportion of natural diamonds are *type Ia*. They contain a fairly large amount of nitrogen (something like 0.3%) which is disseminated through the crystal as aggregates of small numbers of atoms substituting for adjacent atoms of carbon. The aggregations of nitrogen themselves form different types. The stones are usually near-colourless to yellow, with some brown or grey specimens.

Type Ib stones are so rare in nature (perhaps 1 per cent of diamonds) that a yellow type Ib stone is more likely to be a synthetic product. The crystals contain nitrogen, but not so much as in the type Ia stones: the nitrogen here is dispersed throughout the crystal rather than forming aggregates, and substitutes for carbon as individual atoms. Diamonds in this class are usually a deep yellow.

Type IIa diamonds are virtually colourless as a rule, and while nitrogen is believed to be present its amount and arrangement cannot easily be detected. Few natural examples are known.

Type IIb diamonds are also very rare in nature, and appear to contain more boron than nitrogen. They are usually blue or grey, sometimes near-colourless, and will conduct electricity.

While infra-red spectroscopy will distinguish between diamond types, this form of test is well beyond the gemmological context, and the trade needs to rely upon the work of professional geologists and mineralogists to indicate the kind of clues that it needs when a doubtful stone is encountered. While there can be lots of clues, a simple comparison is all that is needed when we want to find out if a diamond is natural, a GE synthetic (doubtful!) or a yellow Sumitomo synthetic.

Under LWUV, type Ia diamonds may show no response or an intense one of orange, yellow, green or blue colour. Type Ib diamonds will show the same colours but with a variable intensity. The GE stones and the Sumitomo crystals are inert. The same types behave in a similar but more variable way under SWUV, while the GE crystals are inert and the Sumitomo stones give a moderate to intense yellow or greenish-yellow fluorescence. Neither GE nor Sumitomo stones phosphoresce, while types Ia and Ib either do show the effect in various colours or do not phosphoresce. This applies to both LWUV and SWUV. Types Ia and Ib phosphoresce with similar variability under X-rays, or do not phosphoresce: GE diamonds do not phosphoresce while Sumitomo crystals give a weak to moderately intense bluish-white phosphorescence. Using the direct-vision (hand) spectroscope, type Ia stones show some sharp absorption bands while type Ib stones show none: neither of the synthetic products shows absorption in the visible region.

When the GIA evaluated the colour of the Sumitomo stones they found that in some specimens it equated to the best natural yellow 'canary' diamonds. While colour zoning was present, this is far from uncommon in natural diamonds: however, in the Sumitomo product there was a deep-yellow inner zone and a narrow, near-colourless outer zone. Within the deeper-yellow coloured zone there could be some subtle colour intensity variation.

Looking inside the Sumitomo stones, all specimens examined by the GIA showed whitish pinpoint inclusions distributed randomly and opaque black metallic inclusions resulting from the flux used in the crystal growth. Such inclusions are not seen in natural diamond. Faceting of the crystals could be carried out in such a way that the inclusions were avoided in the finished stone, since in the crystals they were placed near to the outer edges of the rectangular shape. Vein-like colourless areas could be seen, but again they could be avoided in the polished stones since they extended for only a short distance from the crystal edge. Graining was found to be prominent, with two distinct types observed.

One consisted of sets of lines seen both inside and outside the crystal while the second type, seen only inside the stone, consisted of sets of straight lines radiating outwards from the crystal centre in four wedge-shape formations resembling a cross with splayed ends. Once the stones were polished these effects disappeared: in their place an hourglass shape could be seen through the pavilion of each of the faceted stones examined.

Between crossed polars on the polariscope a cross-shaped interference pattern could be seen: in the crystals it appears most clearly when the rectangular pieces are viewed in a direction parallel to the two parallel polished sides. The effect resembles a bow tie, but when sought in the faceted stones it could not be seen.

Sumitomo stones were slightly magnetic but, as with the GE specimens, the more heavily they were included the more they responded to the pull of a magnet.

Once the dyke of how to grow synthetic gem-quality diamonds was breached the waters flowed in fast! In 1987 diamonds were reported to have been grown by the De Beers Research Laboratory in Johannesburg, South Africa. In the winter 1987 issue of *Gems & Gemology*, the GIA report on the properties of eight crystals and six faceted stones whose weight ranged from 0.27 to 0.90 ct. The stones were transparent and showed no signs of cleavages or fractures: their colour ranged from light greenish yellow through yellow to dark brownish yellow. De Beers stated at the time that growth experiments had been taking place since the 1970s but that work was experimental only and that no crystals had been sold. By 1987, with three firms manufacturing gem-quality diamonds, it was clear that only production and marketing expenses might hinder large-scale introduction of synthetic diamond on to the gem market.

While De Beers were not growing diamond crystals for the gemstone market but rather for research and industrial purposes, the possibility (in 1987) of a large-scale release of material was limited by the cost of production and at the time the company stated that the diamond crystals would be used for specialized industrial applications only. None the less, things could change, and it is worth looking at the properties of the De Beers diamonds.

The brownish-yellow stones showed no response to LWUV, but under SWUV gave a moderate to strong intensity of yellow or greenish-yellow colour with strongly zoned areas which showed no fluorescence. The yellow specimens were inert to both types of UV. The greenish-yellow stones were inert to LW, and gave a weak intensity, yellow-zoned response to SWUV. Only the greenish-yellow diamonds showed any phosphorescence under UV but the effect, a weak intensity yellow, persisted for 10 seconds or longer. No absorption bands could be seen in the visible region in any of the stones.

All stones showed distinct colour zoning with patterns of internal graining, though the greenish-yellow diamonds showed the effect less clearly than the others. Internal graining was also seen with the brownish yellow and yellow stones, producing hourglass-shaped patterns of intersecting lines, an effect less clearly seen in the greenish-yellow specimens, which also gave much less obvious graining.

The inclusion patterns of the De Beers diamonds showed dense clouds of tiny white pinpoints in the brownish-yellow specimens, an effect seen also in the yellow stones where the clouds were less dense. In the greenish-yellow specimens the white pinpoints were isolated and did not form clouds: all three colour types showed metallic inclusions larger than the pinpoint inclusion groups. Faceted stones also showed irregularities or grainy structures on the surface. The irregularities included rough, striated (grooved) and dendritic (plant-like) areas on the girdles.

While most readers will be content to know that the hand spectroscope shows no absorption lines or bands and that therefore other distinguishing features have to be sought, research workers have found that some of the greenish-yellow colour may be due to the presence of nickel during crystal growth: some of the metallic inclusions in the De Beers stones have been found to contain this element. Still on the advanced level, the GIA report that infra-red spectroscopy shows that the De Beers stones contain nitrogen dispersed as single atoms (see above) and that they can therefore be classed as type Ib. Most natural yellow gem diamonds are type Ia (yellow type Ib diamonds are rare), but the type Ib stones also contain some small nitrogen aggregates (characteristic of type Ia diamonds) so that the infra-red spectra of natural type Ib stones also shows features of type Ia material. The two types of spectrum are not seen together in synthetic diamonds since the type Ia spectrum is not present.

None of the De Beers stones were found to conduct electricity, and all showed high thermal conductivity like the GE and Sumitomo diamonds. They cannot be distinguished from natural diamonds by these properties. As with the other synthetic diamonds, the SG was in the normal diamond range. The hourglass strain pattern already noted in the Sumitomo diamonds could also be seen in the De Beers material, and is usually centred in the middle of the crystal. A cross-hatched internal strain pattern is also characteristic.

The De Beers diamonds were found to be weakly to strongly magnetic: few natural stones show this effect to any extent. Apart from the UV fluorescence where the De Beers material shows more varied effects, it does not greatly differ from the GE and Sumitomo stones of similar colours.

In 1993 De Beers produced some boron-doped synthetic diamonds grown for research purposes. The spring 1993 issue of *Gems & Gemology* includes a description of the stones: the sample studied included three rounded modified brilliants coloured light bluish greenish grey, dark blue and near-colourless, weighing 0.075, 0.063 and 0.049 ct, respectively. The clarity grades were approximately VS1, SI2 and SI1. The simpler gemmological tests included fluorescence under LWUV: the light-bluish stone gave a weak orange response to LWUV with the colour unevenly distributed; there was a weak orange phosphorescence of about 1–5 minutes duration. Under SWUV the same sample gave a slightly greenish-yellow fluorescence of strong intensity and uneven distribution. A very strong phosphorescence of a greenish yellow persisted for 1–5 minutes. The dark blue stone under LWUV gave a slightly yellowish-orange fluorescence of moderate intensity with a strong persistent phosphorescence of the same colour. Under SWUV this stone fluoresced slightly greenish yellow with a very strong yellow persistent phosphorescence. The near-colourless stone was inert to LWUV but fluoresced a slightly greenish yellow under SWUV with a very strong yellow phosphorescence. None of these effects would be expected from a natural diamond.

Under the microscope all three stones showed distinct internal growth sectors and pinpoint inclusions similar to those seen in other synthetic diamonds. Metallic flux inclusions were also present. Strong colour zoning was also present, and strain birefringence could be seen between crossed polars. The samples also showed mixed type IIa, IIb and Ib character, a mixture never reported for a natural diamond.

No absorption bands were observed with the hand spectroscope, and other constants were in the normal range for diamond. Perhaps one of the dangers is that if these stones remain in small sizes, gemmologists and others will not trouble to test them.

By 1993 the synthetic diamonds produced from all sources showed features that served to distinguish them from natural stones, and testing should not be too difficult, provided that the possibility of an 'intruder' is borne in mind.

Quickly following the publication of reports on the De Beers boron-doped synthetic diamonds came the Russian gem-quality yellow synthetic diamond, described in the winter 1993 issue of *Gems & Gemology*. The crystals were grown for use in jewellery, and at the time of writing a proposed initiative between Tom Chatham (of Chatham emerald fame) and Russian growers is being discussed: the firm of Chatham was to market diamonds grown in Russia.

The GIA reported on five faceted stones with a yellow to orange or brownish-yellow colour and three supposedly treated yellow to greenish diamonds. The diamonds were grown in Novosibirsk, and the three treated stones were believed to have been heat treated at high pressure after growth. The GIA believe that the Russian diamonds are easy to identify, but again you have to suspect any diamond from the start.

With synthetic diamonds, the response to UV is important since in general it has not been echoed by natural diamond. In this case the non-treated cut stones show a greenish-yellow to yellow fluorescence under LWUV, varying from a weak to a strong intensity but with no phosphorescence. Under SWUV the same specimens responded with a weak to strong intensity yellowish-green to green with no phosphorescence. The treated stones gave a very strong greenish-yellow response to LWUV with a moderate to strong yellow phosphorescence. Under SWUV the fluorescence was strong yellowish green with a moderate to strong yellow phosphorescence.

With the hand-held spectroscope some of the Russian stones showed some absorption features when the specimens had been cooled with a spray refrigerant. The non-treated stones showed a sharp absorption band at 658 nm with a weaker sharp band at 637 nm in one stone and a band at 527 nm in another. The treated stones, also cooled with a spray refrigerant, showed a number of absorption bands between 600 and 470 nm, with less absorption below 450 nm.

The heat-treated stones also show two absorption bands, one with a maximum at approximately 425 nm and the other at 400 nm. There is also a broad band with a maximum at 700 nm, best seen in the greenish-yellow samples. The stones show a number of sharp bands in the visible, including a series covering the range 660–460 nm and a single sharp band at approximately 494 nm.

A number of the samples showed an unevenly distributed weak to moderately intense green luminescence when exposed to a strong source of visible light. This has not been reported for natural stones of this colour.

Testing by infra-red absorption spectroscopy showed that both type Ia and Ib features were combined in the specimens: other synthetic yellow gem-quality diamonds have so far been pure type Ib.

Summarizing these findings we see that the Russian synthetics were the first synthetic diamonds to show a fluorescent reponse to LWUV and that the heat-treated specimens fluoresced more strongly under LWUV radiation than when under SWUV. All the heat-treated diamonds showed a yellow phosphorescence, so that the response of a diamond to UV can no longer be used safely to distinguish between natural and synthetic material. The absorption bands seen in the visible region between approximately 658 and 637 nm and between 560 and 460 nm are peculiar to Russian synthetic diamonds. Some of the bands can be seen with the hand spectroscope. Over the sample considerable variability was noticed, to a greater extent than with previously examined samples of synthetic gem-quality stones. This suggests that during the growth process slightly different conditions of temperature and pressure, with other factors, are used, no doubt to further research.

Though we have so far looked at mostly yellow with a few blue and some near-colourless synthetic diamonds, it would be wrong to expect only these colours in trade conditions. In the fall 1993 issue of *Gems & Gemology* the GIA report on two red diamonds encountered in the trade and found to be treated synthetic diamond.

A dark brownish orange–red stone of 0.55 ct was sent to the GIA's New York laboratory for a report on the origin of the colour. Later in the same year (1993) a stone of 0.43 ct with a dark brownish-red colour was also submitted for an origin of colour report. Natural red diamonds are a very great rarity, and considerable attention was paid to the stones by the laboratory. Gemmological testing showed that the stones were both synthetic diamonds.

The two stones showed very distinct colour zoning, and through the crown facets of the 0.55 ct stone outlines of both square-shaped and superimposed cross-shaped light-yellow areas surrounded by much larger areas of red coloration could be seen. The cross-shaped pattern was more or less central under the table facet. Viewed through the pavilion facets the colour zoning was clearly seen at four positions around the girdle, showing as narrow light-yellow zones surrounded by larger red areas. The 0.43 ct stone showed similar colour zoning through the crown and pavilion.

With reflected light only one graining line could be seen on the table facet of the 0.55 ct stone. The other specimen showed a faint surface graining pattern on the table facet with some parallel polishing lines. Inside the stones were large opaque metallic-looking inclusions, and both the diamonds were attracted by a pocket magnet.

Under UV radiation of both types the crown facets of the 0.55 ct stone showed unevenly distributed very intense green and moderately intense reddish-orange regions: the green fluorescence corresponded to the light yellow zones described above. Under SWUV these sections gave a phosphorescence lasting for several seconds. The reddish-orange fluorescence was seen only at one small point near the girdle under LWUV, but under SWUV this colour was seen on all the large dark-red areas. The 0.43 ct stone also showed uneven luminescence but not in the same pattern. Looking through the crown facets under either type of UV a very small area of red fluorescence could be seen near the centre of the table facet, this area being surrounded by a narrow zone of green fluorescence. Narrow bands of an orange fluorescence pointed from the green-fluorescing area towards the four corners of the table facet where there were areas of stronger orange fluorescence. The rest of the stone fluoresced a weaker orange-red, and there was no phosphorescence. When the 0.55 ct stone was illuminated by strong visible light a moderate green luminescence could be seen in the yellow areas. This effect was not observed in the other specimen.

A number of sharp absorption bands could be seen between 800–400 nm in the spectrum of the 0.55 ct stone. Some between 660 and 500 nm were visible with the hand spectroscope and were seen best when the stone was cooled. Fewer absorption bands were seen in the 0.43 ct stone though they occupied similar positions. Both stones showed increasing absorption towards the violet, and a broad absorption region extending from 640 to 500 nm. Using infra-red spectroscopy the stones were shown to contain elements of both type Ia and Ib diamonds (they can be more closely classified).

Compared to other synthetic diamonds the two stones show what appears to be a unique fluorescence, with an especially uncommon response to LWUV. Study of the various spectra shows that nickel is present in the diamonds, which were probably grown in a nickel-containing flux. Some of the bands in the visible spectrum of the 0.55 ct stone indicate that irradiation and heating had taken place.

Compared to natural diamonds the two red stones show several distinguishing features. Although some type IIa diamonds of natural pink colour show an orange fluorescence under UV, the GIA had never seen a known treated pink IIa diamond: some of the pink to red diamonds whose colour is natural show a blue fluorescence and are classed as type Ia. Some pink to red treated natural diamonds of type Ib show the absorption bands at 637, 595 and 575 nm that are characteristic of treatment by irradiation and heating. Neither they nor two treated natural diamonds of mixed types Ib and Ia with a pink to yellow colour and moderately strong orange fluorescence to LWUV and SWUV, resembled the two diamonds examined, as they did not show the pattern of colour zoning nor the absorption bands related to nickel.

In 1990 GE Research and Development Center in Schenectady, New York, announced that they had manufactured near-colourless, isotopically pure carbon-12 synthetic diamonds. In nature, carbon is a mixture of carbon atoms of different weights (isotopes): while nearly all carbon has 12 atomic mass units (12C) some has 13 (13C). With the technique of isotope enrichment it is possible to make carbon of either type in a nearly pure state. While the manufacture is expensive, diamonds of isotopic purity have many technological applications.

At the time of the paper published in the fall 1993 issue of *Gems & Gemology* no faceted diamonds of this type were known. However, examination of the crystals showed rod-like inclusions of metallic appearance and clouds of tiny triangular or lozenge-shaped tabular inclusions in areas beneath the octahedral crystal faces with tiny pinpoint inclusions randomly scattered throughout. These had a bright, white metallic appearance in reflected light and looked brownish in transmitted light. No graining could be seen in either of the crystals but strain (anomalous birefringence) was seen between crossed polars as a weak pattern of grey or blue. In general, synthetic diamonds rarely show indications of strain.

Both the crystals remained inert to LWUV but fluoresced a weak yellowish orange under SWUV. This colour seemed to show an increase in intensity when the radiation was turned on, then remained at a fixed level.

More important was the cathodoluminescence shown by the crystals. This effect is obtained by subjecting the specimen to a beam of electrons in a vacuum chamber. In the two crystals a zoned pattern of cathodoluminescence could be seen, the pattern corresponding to the arrangement of different internal growth sectors. The colour produced by the electron beam was a slightly greenish blue. Under X-rays the crystals gave a yellow luminescence with very persistent yellow phosphorescence, one crystal displaying the effect for at least 10 minutes. No electroconductivity was observed in either crystal nor were there any absorption bands seen in the visible spectrum.

Compared to earlier GE synthetic diamond crystals, the two crystals under discussion, with no electroconductive powers, show that they belong to type IIa compared to the earlier type IIa/IIb GE crystals which are electroconductive. This is thought to be due to some change in the growth process rather than to their isotopic composition. As always, compared to natural diamonds, the new GE crystals show strong fluorescence under SWUV and metallic inclusions.

Summary of synthetic colourless or near-colourless diamonds

Up to the present, near-colourless diamonds should be checked for a response to SWUV, for the presence of metallic inclusions and for magnetism when these are prominent. Anomalous birefringence between crossed polars is more likely to be seen in natural diamonds and synthetic diamonds usually show Cape absorption bands. Signs of natural inclusions should be looked for first in any diamond.

With the winter 1995 issue of *Gems & Gemology* a chart for the separation of natural and synthetic diamonds was included. Some of its features as they affect synthetic diamond are given below.

Synthetic *colourless or near-colourless diamonds* will be type IIa or mixed type IIa + Ib + IIb. They may appear colourless or light grey or very light blue, yellow or green. Some of the colour may appear in sectors. Cut stones so far have not exceeded 1 ct in weight.

Between crossed polars a black cross effect may be seen due to anomalous double refraction. Metallic inclusions may be seen by reflected light – they appear black by transmitted light. They may be long or rounded and occur singly or in groups. In addition, clouds of pinpoint inclusions may be seen.

Generally the colourless to near colourless synthetic diamonds show no response to LWUV. They are much more likely to show yellow, greenish-yellow or orange–yellow weak to strong fluorescence under SWUV. Where fluorescence is seen, its distribution may be uneven, sometimes taking a square, octagonal or cross-shaped pattern. So far, colourless and near-colourless synthetic diamonds have always shown phosphorescence after irradiation by SWUV. The effect may persist for at least one minute, and the colour is usually yellow or greenish yellow. Synthetic diamonds of this type show no fluorescent response to visible light. No sharp absorption bands are detectable with the spectroscope.

Some specimens may be attracted by a small magnet and some show electro-conductivity.

Synthetic *blue diamonds* are type IIb or mixed type IIb + IIa. The colour ranges from light to dark blue with some specimens showing a greenish or greyish blue when small yellow growth sectors are present. Magnification shows internal growth sectors, some with a central octagonal shape surrounded by octahedral and cubic faces. Growth sectors show up well under UV. An even colour distribution is also noticeable with some variations seen distinctly and others less perceptible. Graining planes intersecting in patterns are also characteristic. Hourglass shapes are often seen, and weak, cross-shaped anomalous double refraction shows between crossed polars.

Inside synthetic blue diamonds opaque black metallic inclusions can be seen as in the colourless and near-colourless stones. Clouds of tiny pinpoint inclusions have also been observed.

Under LWUV the synthetic blue diamonds are inert though they respond to SWUV with a yellow or greenish-yellow fluorescence. As usual with synthetic diamond the response to SWUV is stronger than that (if any) to LWUV. The fluorescent effects seen are unevenly distributed, and duplicate internal growth sectoring with octagonal or square patterns. Some sectors may show fluorescence while others remain inert. There is usually a yellow moderate to strong persistent phosphorescence which may last up to one minute. Cathodoluminescence shows a distributed effect.

There is no luminescence to visible light nor any detectable absorption spectrum when the hand spectroscope is used. Stones are electrically conductive and may be attracted by a strong magnet.

Synthetic yellow diamonds may be type Ib or Ib + Ia. They are Ib or IaA if irradiated and heat-treated or type IaA if heat treated at high pressure. (The class IaA and other finer classifications are explained elsewhere.) The colour range is from greenish yellow through orange yellow to a brownish yellow. Type Ib or IaA treated stones show an orange to pink or sometimes red colour (after irradiation and heating to about 800°C) while IaA treated stones heated from 1700 to 2100°C at high pressure show a yellow to greenish-yellow to yellowish-green colour.

Looking at colour distribution, growth sectors can be seen inside the stones, and are often octagon-shaped with additional cubic sectors. The colours usually form dark and light yellow sectors, while stones which have been irradiated and heated to give a pink or red colour show pink or red zoning with some yellow sectors.

Graining planes can be seen between internal growth sectors, and the planes often intersect to form patterns. Hourglass or other effects can be seen with transmitted light. Between crossed polars a black cross may be seen: this is due to anomalous birefringence.

Inclusions of flux metal can usually be seen with tiny pinpoint inclusions: some of the flux metal inclusions may reach 1 mm in length.

Under LWUV, type Ib stones show no visible response while type Ib + IaA stones show weak to strong yellow or yellow–green fluorescence. Type Ib or IaA treated diamonds show a strong green plus a weak orange fluorescence in different growth sectors while in some stones only a weak orange colour is seen. Type IaA treated stones show a greenish-yellow or yellow fluorescence which is very strong. Under SWUV, type Ib stones show either yellow to yellow–green fluorescence of weak to moderate strength while type Ib + IaA stones give a yellow–green response which may range from weak to strong. Type Ib or IaA treated yellow diamonds give a strong to very strong green plus a weak orange fluorescence in different internal growth sectors: some stones merely give an orange fluorescence. Type IaA treated stones give a strong greenish-yellow colour. The intensity of the fluorescence varies: with type Ib or Ib + IaA stones it is stronger under SWUV than under LWUV; with Ib or IaA treated stones the response is either of equal strength or stronger under SWUV. With IaA treated stones the intensity is stronger under LWUV.

The fluorescent colours are unevenly distributed and duplicate the growth sector arrangement. With type Ib or Ib + IaA stones there is usually no phosphorescence, but a weak yellow or greenish-yellow effect may be seen, persisting for some seconds. With type Ib or IaA treated stones there may be a weak orange phosphorescence for several seconds, while with type IaA treated diamonds there may be a strong and persistent yellow phosphorescence lasting up to one minute.

The only luminescence from visible light may be seen in Ib + IaA stones with a weak to moderate green colour. Other types may show a similar effect, sometimes of a weak orange colour. Type Ib stones do not respond to visible light.

In type Ib stones no sharp absorption bands are usually seen, though if the stone is cooled the 658 nm band is sometimes visible. In type Ib + IaA stones several sharp bands may be seen if the stone is cooled: they are at 691, 671, 658, 649, 647, 637, 627 and 617 nm. In type Ib or IaA treated stones several sharp bands may be seen at 658, 637, 617, 595, 575, 553, 527 and 503 nm. Type IaA treated stones may show sharp absorption bands at 553, 547, 527, 518, 511, 503, 481, 478 and 473 nm. Stones may be attracted by a strong magnet.

Diamonds with enhanced colour

While fine colourless diamonds have always commanded high prices, diamond merchants over the years have had to accept that many 'colourless' specimens are nowhere near this quality. The Cape stones in particular show a highly characteristic off-white to pale-yellow colour (nothing like the canary yellow of a fancy stone) which is not too attractive. It is hardly surprising that attempts are made to obtain more fancy stones, and if they can be achieved by some form of permanent alteration of stones whose colour makes them a poor sales prospect, while being otherwise clean, undamaged and of fair size, so much the better.

Some early attempts to improve the colour of diamonds involved the use of radium. Crookes buried diamonds in radium salts for periods of up to one year, and found that while the colour changed to green (presumably from a Cape colour) the stones became radioactive. Nassau (1994) reports that the radioactivity arises from the implantation of fast-moving nuclei recoiling from their disintegration and entering the diamond surface. Very high levels of radioactivity are found in some of these green stones; figures up to 80 mrem/h have been recorded. In some stones spots on the pavilion surface show radiation localization, and although high temperatures may alter the green colour (produced by α-particles) they do not remove the radioactivity. Americium is reported to have been used with the same coloration and effects.

Several techniques have been used to alter the colour of diamond by irradiation. They include neutrons from a nuclear reactor, α-particles, protons and deuterons. To the gemmologist the source of the radiation is immaterial since he or she needs to know only how to test a suspected stone. The sources just listed produce an absorption band which can be seen with the hand spectroscope: extending from the infra-red into the yellow–green region, its presence gives the diamond various dark shades of green or even black. That part of the band occupying part of the infra-red cannot of course be seen with the hand spectroscope, but its importance is such that it is always known as the GR1 band ('GR' denoting general radiation).

Because the sources listed above are not very powerful, the colour they give to the diamond does not penetrate the specimen completely. We shall see from the reports at the end of the chapter that laboratories note a characteristic umbrella-like marking above the culet in the pavilion of the stone. This is not difficult to see, and is diagnostic for an irradiated diamond. This shows that irradiation took place from the pavilion side: diamonds irradiated through the table show dark patches in different positions, and if the stone is placed table facet down on a white background a dark ring will be seen.

Another method of treating diamond is by high-energy electrons. During this treatment heat is developed which could affect the final colour produced: if the heat is not wanted the stone needs to be cooled during treatment. While the gemmologist will find that electron-treated stones have uniform coloration of blue to blue–green, identification is not as easy as it is for diamonds treated with heavy particles. Depending upon the energy used, the colour may become blue with low-energy electrons and a more greenish colour when the energies are raised. The GR1 band makes its appearance in electron-treated stones.

It is clear that if you are going to irradiate a stone you are most likely to choose a process which at least colours the stone right through. There is no problem with discoloration or fading with any of the diamond irradiation processes, and stones do not usually become radioactive because impurities in diamond (whose presence makes the stone more likely to become radioactive) are generally low. Nassau (1994) describes a black diamond (in reality a very dark green) in which metallic polishing residues in surface-reaching fractures were activated by irradiation. Nassau recounts how boiling in acid removed the radioactivity, which passed to the acid!

We shall see from the laboratory reports at the end of the chapter that irradiated diamonds at times show unusual (and probably unexpected) colours. Nassau (1994) reports a stone which before irradiation fluoresced a greenish-yellow and which was expected to become a chartreuse colour after treatment. In fact the diamond, treated by neutron irradiation, became orange–red and gave a bright orange–red fluorescence. The stone may have possessed type Ib characteristics. Another stone reported by Nassau was also neutron irradiated: cloudy before treatment, it became a rich sky-blue colour. Nassau asks the question which many must have posed to themselves – can you make a natural blue diamond (type IIb semiconducting) a darker blue by irradiation? If you could, the stone might cause problems because a blue electroconductive stone might well be classed as natural should a conductivity test be the first one to be applied. Nassau (1994) says that one stone lost its conductivity after irradiation.

When a diamond has been irradiated by whatever method the resulting colour is usually too dark and unattractive (dark green, dark greenish blue) to leave the treatment process at that point. Heating (the term 'annealing' is often used) develops a more acceptable colour. The dark irradiation-produced colours, on heating, turn from dark blue or dark green to brown and then to yellow, returning finally to the pretreatment colour. When conditions are appropriate the heating stops and the stone is removed for sale.

While yellow diamonds are the most common product of irradiation and heating, other colours are often reported. When a pink or red colour results, the original diamond will have been type Ib. Such stones are reported by Crowningshield (*via* Nassau, 1994) to show a strong orange fluorescence. Natural pink diamonds also possess this property.

In *Nature*, Vol. 273, p. 654 (1978) Collins describes how heating from about 400°C upwards destroys the GR1 band and either forms or intensifies bands at 595 nm (sometimes the values 592 or 594 nm are quoted), 503 nm and 497 nm. This absorption pattern gives the diamond colours: dependent upon the intensity of the heating, orange, yellow, green or brown may be developed from the blue created by the irradiation. If a treated diamond is heated to 1000°C the band at 595 nm will disappear though the colour does not change, so a gemmologist encountering a yellow diamond which does not show this band cannot safely say that the stone has not been treated. In natural yellow diamonds the band at 504 nm is usually easier to see than the band at 497 nm: after irradiation and heating, the relative strengths of the two bands may be reversed or at least equal. Cape absorption bands should lurk in the stones, however, and can usually be brought to visibility by cooling the specimen.

Some years ago in a conversation, Frederick H. Pough told me that he had personally been responsible for the irradiation of a celebrated diamond, which had started life as the 104.88 ct Deepdene. He published the admission later in the *Lapidary Journal*, Vol. 41(12) (1988). The stone was originally a Cape colour and was turned to a deep green by irradiation: heating then turned it to golden yellow. In 1971 the stone was offered for sale as a natural yellow diamond, but its identity was recognized even though the weight was now slightly less at 104.52 ct.

Some type Ia diamonds may change to a bright yellow on heating to near 2000°C for minutes only (Field, *The Properties of Diamond*, 1979, Academic Press) and some polishing processes have developed enough heat to cause a colour change in diamond. Sometimes a fancy light yellow can develop in conjunction with the absorption band at 503 nm. At the time of writing there is still plenty of speculation about colour development in diamond and how it may be recognized.

Diamonds with enhanced clarity

The standard method of grading polished diamonds is by consideration of colour (grades of colourless), cut and clarity. Clarity is assessed by the presence or absence of visible inclusions and, when inclusions can be seen, by their size, colour and position within the polished stone. Despite the unique hardness of diamond, fractures are not uncommon, and some of them may result from cleavage (diamond breaks cleanly along certain crystallographic directions). When fractures are profuse and reach the surface, it is not surprising that attempts to conceal them in some way have been made for many years since the consequent improvement in clarity grade makes the stones more valuable.

In the 1980s, ways were found in which surface-reaching fractures in diamond could be filled with substances which would make them less visible and so enhance their clarity grade. One of the first to develop one method of fracture filling was Zvi Yehuda of Ramat Gan, Israel. The manufacturer reports that the diamonds are cleaned and then filled with a molten glass. The filling takes place at high temperatures, and any glass remaining on the surface is removed. Since 1980, fracture filling has developed to become commonplace for both the colourless and fancy colour stones coming on to the market.

Size is no barrier since reports of both small and large stones occur in the literature: a report in the fall 1994 issue of *Gems & Gemology* cites filled diamonds of 0.02 ct and one publicity report quotes specimens of 50 ct. At the time of writing, the whole topic of fracture filling is beset with claims and counter-claims about how durable the different and rival processes are, how easily the practice can be detected and what is the position on disclosure when fillings are detected at the sales point.

The best test for any kind of treatment is *suspicion*. Without this the stones will never be tested at all since there is never time. Diamond is a common if expensive gemstone, and even the smallest retail stock will be far too large for every stone to be tested as a matter of routine.

Once suspicion has been aroused (or you have a particularly expensive diamond) the 10× lens and the microscope are the instruments to call into the hunt. Try all kinds of lighting – you will find a fibre-optic source essential since you can move the beam of light around to suit your purpose. Reflected and transmitted light will be provided by your microscope. While gemmologists may want to find out the nature of the filling used, the jeweller only needs to know that it is there: details of some of the fillings are given here when their nature affects testing, and others in the report section at the end of the chapter. The durability of the materials used for filling is naturally an important question but this again is a matter for the laboratory.

As mentioned below, lasers have been used to remove unsightly inclusions.

A diamond-filling experiment

In the fall 1994 issue of *Gems & Gemology*, some filled diamonds are highlighted, and the Israel-based firm of Koss & Schechter Diamonds supplied specimens which they had attempted to fill, using two different processes. One mirrored the processes in general commercial use and is based on halogen glasses while the other is experimental and based on halogen oxide glasses.

With the Koss stones, orange and yellow flashes could be seen inside the filled stones when rotated under dark-field illumination. Under bright-field lighting, flashes of blue and violet were seen. These colours are not see in all filled diamonds. All filled areas contained gas bubbles, and some of the stones tested showed flow structures in the fillings. Some of the filling material was distinctly yellow, but fine crackled lines seen in some other stones treated by the same firm were not seen. All stones were found to contain lead and bromine in the fillings by energy-dispersive X-ray fluorescence analysis (not available to the general gemmologist).

Stones treated with halogen oxide glasses did not fill successfully though lead and bromine were found in the small amounts of filling material that did enter one of the specimens.

While at one time Koss & Schechter claimed to be about to incorporate a fluorescent additive in their filler, stones with this effect have not (1996) so far been reported, nor did the use of cathodoluminescence show up any additives. Examining the unsuccessfully filled diamonds, however, did show up a wide range of flash colours, including red, orange, yellow, blue, purple and pink. These colours were seen in dark-field conditions: under bright-field lighting, bluish-green, green and greenish-yellow colours were seen.

These are valuable first-hand reports, and the firm's cooperation is welcomed by gemmologists and the gemstone trade. An earlier gemmological study in the summer 1989 issue of *Gems & Gemology* looked at diamonds treated by Mr Yehuda. Working as always on the principle of using a filler with an RI as close as possible to that of diamond (like ice in water, this means that the filled area will be hard to see), Yehuda is reported (though not proved) to have used relatively high pressures in his filling process, with temperatures in the region of 400°C. It is possible that a vacuum was used.

Diamonds selected from the GIA collection were sent to Dialase Inc. of New York, a firm advertising a filling process developed by Yehuda. Six round brilliants were sent and tested on their return. Grading for clarity and colour took place before and after treatment. Results over the six stones were very varied, some specimens looking much better while others seemed virtually unchanged. Answering the question of weight addition through filling, the stones showed no sign of weight gain: this is presumably because the filling takes the form of a thin film. Interestingly, four out of the six stones showed a fall of one full colour grade after treatment, and two stones showing no significant improvement in clarity also dropped to a lower colour grade.

Looking at the stones under the microscope the filled specimens had a slightly greasy or oily appearance with a very slight yellowish tinge. With the lens, diamond dealers would suspect a stone with a large number of surface fracture signs. The Dialase-treated stones showed the flash effect, with a characteristic yellowish-orange colour seen under dark-field lighting. This changed to an intensely vivid electric blue under bright-field illumination. Tilting the stone backwards and forwards will show the flash colours changing from orange to blue to orange. The colours are seen best when the stone is viewed at a steep angle and close to a direction parallel to the plane of the treated fracture. The flash effects are not always easy to see: their visibility depends upon the colour of the stone, so that in a dark brown diamond only the blue flash colour will be apparent. In very small filled fractures the flash effect may not be seen.

The orange flash colour should not be mistaken for the quite common iron-staining colour seen in some untreated diamonds, and some unfilled fractures acting as thin films may produce interference colours: they will normally show more colours together than the single orange or blue flash colour.

In the fillings, a flow structure may be seen and gas bubbles are often visible: these may resemble the fingerprint inclusions seen in other gemstones though not in diamond. When the filling is examined it often shows a light-brown, light-yellow or orange–yellow colour.

The nature of the filling material used in the Dialase stones has been found to be a compound of lead, chlorine and oxygen with variable amounts of bismuth and perhaps boron, which is hard to detect. The *Gems & Gemology* report of 1989 suggested that a lead–bismuth oxychloride (perhaps similar to the natural mineral perite) may have been used. This material has an RI close to that of diamond so that it would be hard to detect without the flash colours.

The Dialase filling was tested for stability and found to be unaffected by ultrasonic cleaning, by steam treatment or by boiling in a detergent solution: the filling was unaffected by thermal shock and by stress in the setting process. Repolishing damaged the filling in some instances, and heat from a midget torch used in repair was found to cause beads of the filling agent to sweat out and appear on the surface when the diamond had been heated nine times.

It is not possible to state too many times that filling detection starts from a suspicious attitude to as many diamonds as possible. When a filled stone is examined the process should show up without too much difficulty.

The Dialase diamonds filled by the Yehuda process, and stones treated by Koss & Schechter and by Clarity Enhanced Diamond House, all described in the 1994 *Gems & Gemology* paper, conform to the stones described in the earlier 1989 paper – flash effects continue to be the best clue to fracture filling.

In 1995 a chart published with the summer issue of *Gems & Gemology* showed 56 colour photographs illustrating the main features of filled diamonds, under the headings flash effects, flow structure, trapped bubbles, misleading features and testing techniques. The misleading features include such points as interference colours, mentioned above, a feathery appearance in unfilled breaks, natural iron staining of brown or orange, brown radiation staining and burn marks on a diamond surface (from the polishing process), which last may be mistaken for the remains of the filling substance left on the stone after treatment.

The full diamond classification

The difference between type I and type II diamonds is the presence of nitrogen in type I and its absence in type II. Type Ia diamonds (more than 98 per cent of large clear natural stones) contain nitrogen either as pairs of atoms (type IaA) or as larger clusters containing an even number of nitrogen atoms (type IaB). Type Ib stones are very rare, containing isolated nitrogen atoms (this type includes less than 0.1 per cent of known stones).

All type I stones absorb in the infra-red region from 6 to 13 µm, and in the UV beyond 300 nm. They usually show a blue fluorescence and the yellowish Cape stones belong to this type. The characteristic absorption band at 415.5 nm (seen in stones which began as Cape diamonds and were then treated to improve their colour) is probably due to three nitrogen atoms surrounding one atom of carbon in a flat configuration: the band is known as the N3 band. Other less prominent bands can be seen.

Type I diamonds with a brown rather than a yellowish body colour show a weak absorption band at 504 nm, and some specimens fluoresce a greenish colour. No type I stones conduct electricity.

Type II diamonds contain no nitrogen, do not absorb in the region 6–13 µm and transmit in the UV below about 300 nm. They are good conductors of heat. Type IIa diamonds are colourless and contain virtually no impurities, transmitting in the UV down to near 250 nm. Type IIb diamonds contain boron and are usually blue: they are rare and include only about 1 per cent of all diamonds. After exposure to SWUV they may give a blue phosphorescence, and some diamonds of this type have a grey or blue body colour. Type IIb stones transmit down to about 250 nm and are electro-conductive. The blue stones do not show a visible absorption spectrum.

Diamond thin films

These are most conveniently studied under the heading of diamond since in principle they may be applied to any gem species. The thin film is composed of synthetic diamond, of course, and is approximately 1 µm (0.001mm). Growth takes place on a substrate, and the temperatures and pressures required are not particularly high.

Crystal growers are familiar with the technique of chemical vapour deposition (CVD), and the deposition of diamond thin films on a gemstone host makes use of some of the techniques of this process designed for larger crystals. It has generally been considered that the growth of diamond thin films is easier and cheaper than the growth of a synthetic diamond crystal. The point of applying a diamond coating is to improve the resistance to abrasion of the coated substance, and in the case of gemstones to improve their appearance.

While diamond and gemstone dealers need not expect to encounter diamond-coated stones in any quantity at the time of writing, several features make identification fairly straightforward. The presence of interference colours on the surface is accounted for by the difference in RI between the substrate and the diamond coating: it may also be seen when the diamond coating is not in direct contact with the surface, a film of air lying between.

The diamond films have been found to give a hazy appearance: this is best seen under dark-field illumination when the coating seems less transparent than the effect expected from a non-coated stone. The haziness is due to light scattering within the coating, whose nature is polycrystalline. If the coated stone is examined in diffused light and held against a white background, the film makes the stone look brownish on the surface. No absorption features can be seen in the visible region. Seen between crossed polars, a coated stone shows no extinction in the film, again due to the polycrystalline nature of the coating. Testing for thermal conductivity has given similar readings for the diamond coating and for a silicon substrate, so coating a non-diamond with this type of film would not conceal its true nature. In general the thermal conductivity tester gives the best results for a diamond thin film on a gemstone substrate. It has been calculated that a diamond thin film would have to be at least 5 μm thick for the coated stone to pass as diamond when it was another species.

It is possible that a thin film coating might be used to alter the colour of a polished diamond: Sumitomo Electric Industries reported that a blue film of 20 μm had been deposited on a natural near-colourless diamond octahedron, giving a blue crystal which was electro-conductive. Immersion of the stone might show this kind of coating since the colour of natural blue diamond is usually patchy and the film would show sharp edges. It is possible that a thin blue coating might go some way towards improving the colour of a Cape diamond substrate.

While coating by CVD is a fairly new technique as far as gemstones are concerned, the simpler practice of blooming camera lenses to minimize the effect of reflections has been known for many years. Again, the idea is to make a yellowish diamond look whiter, and if the coating is not noticed when turning the specimen in a strong light (when the bluish bloom will be seen), the presence of a spotty or granular area near the girdle, with a pitted appearance, should give the game away. It has been found that some coatings can be removed by boiling in sulphuric acid, often showing the original yellowish colour of the diamond beneath.

On occasion, diamond crystals have been burnt with the intention of oxidizing the surface and giving a whitish appearance. In this way a yellowish crystal may look like a white one with a frosted surface. It is believed that a polisher once acquired a parcel of such crystals, which then, on cutting, resumed their yellow colour.

Summary

Glass takes pride of place as the commonest if not the most convincing imitation of diamond. Showing no crystalline properties it is soft and fractures easily, is a poor conductor of heat, contains well-rounded and prominent gas bubbles and shows a swirly internal structure. A piece of jewellery containing many matching coloured 'diamonds' is very likely to be glass. Glass contains no natural mineral inclusions and may show rounded facet edges when the stone has been moulded rather than faceted. On the refractometer it will very frequently give a single reading in the RI range 1.50–1.70.

All other diamond imitations are crystalline and, apart from natural zircon, which shows high birefringence, contain no mineral inclusions – a very unusual feature for a diamond. Since man-made substances such as YAG, GGG and, most of all, CZ give readings too high for the refractometer, the reflectivity meter or thermal conductivity tester is the best instrument to use for testing. Most of then read as 'diamond/not diamond', but this is all the trade needs to know.

Synthetic spinel can be tested on the refractometer but also gives a sky-blue fluorescence under SWUV: so far only some synthetic diamonds show this property.

Some man-made substances are at their most dangerous when fashioned as the base of a composite with a hard transparent top of some other material. Strontium titanate is clear and colourless with a dispersion ('fire') exceeding that of diamond: set with a crown of synthetic spinel, corundum or CZ the dispersion is muted and even more diamond-like. The diamond doublet shows a reflection of the table facet inside the stone.

Coloured diamonds may be imitated by YAG, GGG or CZ doped with different rare earth elements which may often give a characteristic fine-line absorption spectrum. Synthetic rutile, though highly birefringent and with very high dispersion, often turns up as a convincing simulant, in just those rather dull colours of blue and brownish yellow shown by many diamonds.

UV radiation is very useful in diamond testing though the response of natural stones is less predictable than that of synthetic or treated diamond. It is possible that an unmounted specimen may need an SG test, though such a time-consuming process can usually be avoided.

Synthetic gem-quality diamond in colourless, near-colourless, blue and yellow has been made by more than one manufacturer, and we have seen that specimens are often magnetic owing to their metallic inclusions, may fluoresce under SWUV, show characteristic graining patterns and may combine the characteristics of more than one of the diamond types. Between crossed polars many synthetic diamonds give an hourglass or bow-tie effect. Their electro-conductivity, when possessed, may not echo that of natural diamond. The colours produced by fluorescence or cathodoluminescence often follow growth sectors. Synthetic diamonds may be irradiated and heated (annealed) to give enhanced colours. The gemmologist can identify a synthetic diamond so long as the possibility of one turning up is borne in mind.

Many diamonds are worth more when coloured, and for a long time Cape diamonds of a yellowish, off-white colour have been irradiated and annealed to produce a variety of much more attractive colours, of which a bright yellow is the commonest. Yellow treated diamonds also give the most trouble to the gem tester, and while the spectroscope will often show that the specimen began life as a Cape stone, cryogenic conditions are sometimes needed for this test to work. Other colours can also be identified with the spectroscope.

If fractures in diamond can be filled with a transparent colourless substance with a similar RI to that of diamond, the fracture 'appears to disappear', leaving the stone in a higher-clarity grade. Various substances have been used, and more than one firm has treated diamonds both experimentally and commercially. While the colour grade is sometimes lowered by filling, many diamond dealers accept the practice, the only unresolved difficulty being whether or not to disclose it, when seen, to a customer. Filling material may contain a flow structure and gas bubbles: it may show a yellowish oily or greasy appearance and, most of all, unexpected flashes of orange or blue colour inside the stone, the colour seen depending on the type of lighting used for examination. Unsightly inclusions may be removed or their effect diminished by lasering, though the laser tracks are easy to see.

Plate 1 The classic Verneuil inclusion-curved growth lines in a synthetic ruby. This effect is not always easy to see, and careful positioning of the specimen may be needed

Plate 2 Large gas bubbles with a bold outline are found only in glass or flame-fusion-grown stones, as in this ruby

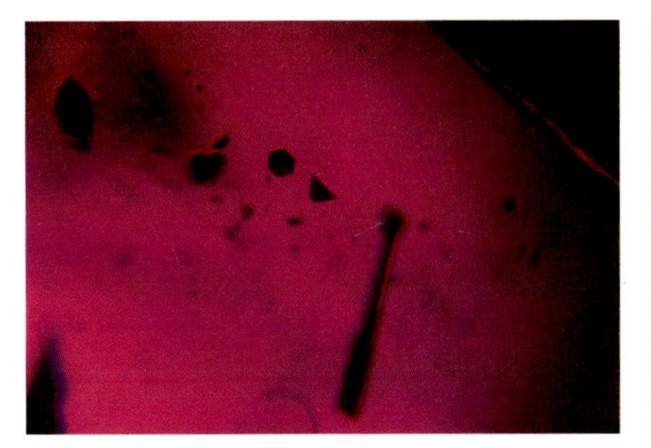

Plate 3 Detached crucible material shows up in a Chatham flux-grown ruby. Angular metallic fragments never occur in natural specimens

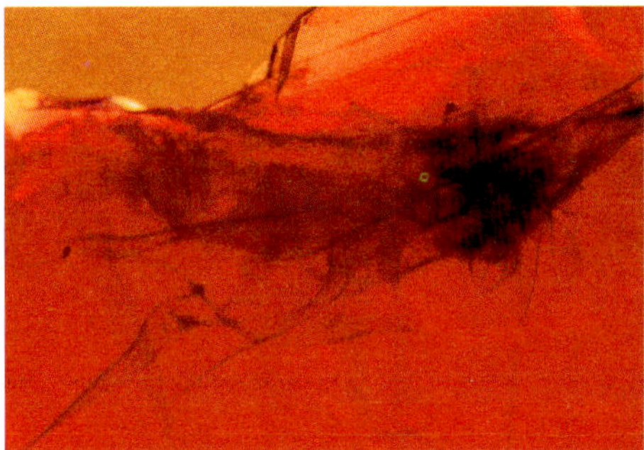

Plate 4 Twisted veils of undigested flux material characterize flux-grown stones. Apparently similar effects in natural stones arise from masses of liquid droplets, and their arrangement is rarely twisted. This is a Ramaura ruby

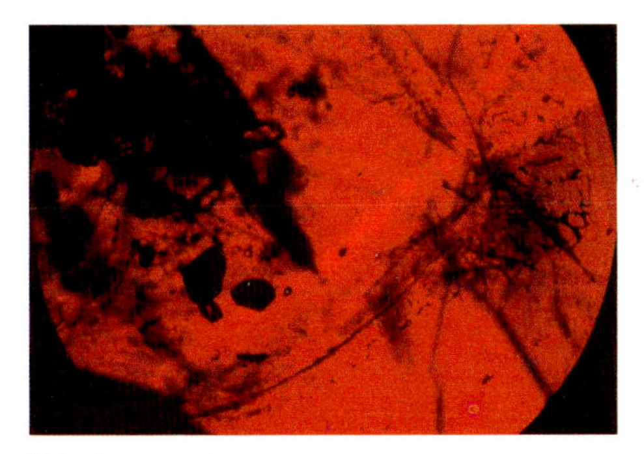

Plate 5 A very deceptive specimen – synthetic overgrowth on a seed of natural Thai ruby

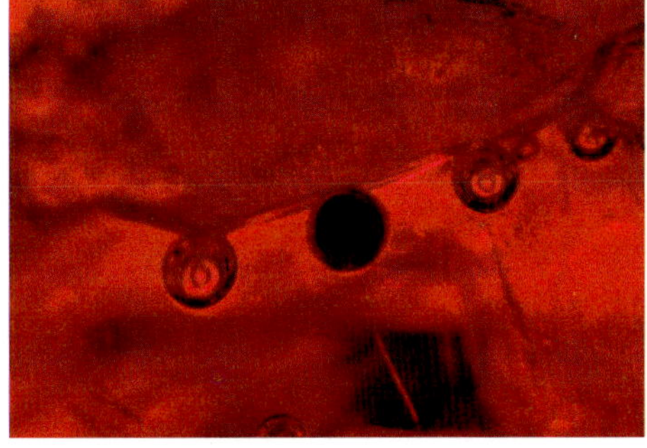

Plate 6 A gas bubble in a natural ruby indicates a glass infill – here the bubble is sectioned

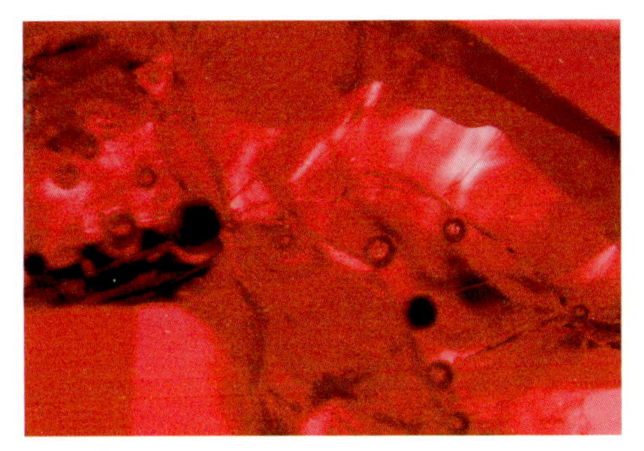

Plate 7 A glass-filled cavity in a ruby

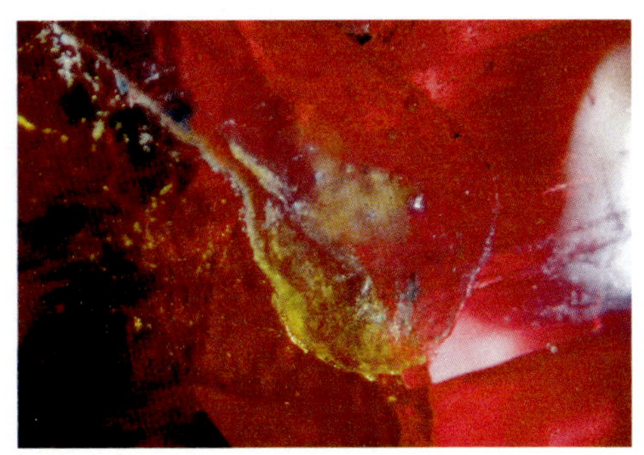

Plate 8 An unexpected brownish tinge betrays a plastic infill in a ruby

Plate 9 Curved growth lines in a Verneuil blue sapphire

Plate 10 Crucible fragments show sharp and black in a Chatham blue sapphire

Plate 11 Bubble trails in a synthetic yellow sapphire

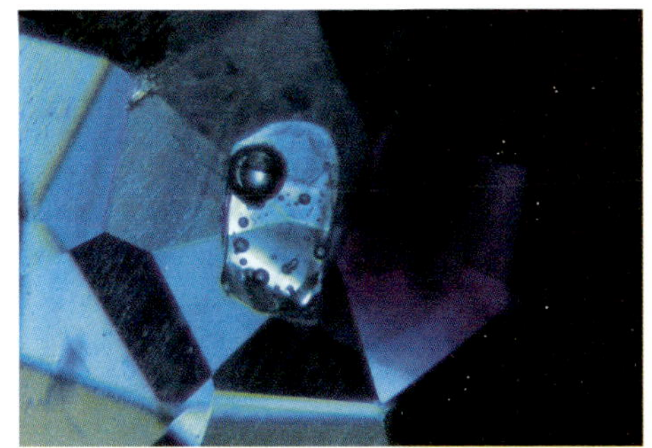

Plate 12 Glass-infilled blue sapphire indicated by a gas bubble in the glass

Some stones are coated with thin films of diamond and show interference colours on the surface.

Reports of interesting and unusual examples from the literature

Items in this section have been chosen to illustrate points made in the chapter and to bring one-off items to your notice.

Coated diamonds seem to appear in the trade at intervals: in the summer 1984 issue of *Gems & Gemology* the GIA report stones which before coating might have been classed as colour grade H or I. After coating the grade might very well rise to G, but the coating seemed always to impart a suspiciously grey appearance. Special lighting techniques described in *Gems & Gemology* for winter 1962 are useful in the detection of coatings. Sometimes a faint band can be seen on one side of the girdle: some diamonds show the band both on the crown and the pavilion side of the girdle.

Though a technique used only by research laboratories, cathodoluminescence can help to distinguish between natural and synthetic diamond crystals. Details of the process and illustrations of some examples are given in the *Journal of Gemmology*, Vol. 24(7) (1995).

In the issue of the *Rapaport Diamond Report* for 4 September 1987, Mr Zvi Yehuda of Israel stated that he introduces a 'secret ingredient' into stones at 50 atm pressure and 400°C temperature: the stones chosen for the treatment are heavily flawed. The *Report* recommended dealers to look for small bubbles around the treated area, using 20× magnification, as well as for the iridescent effect.

Under the heading 'Want to buy a hot diamond?' the summer 1987 issue of *Gems & Gemology* describes a black stone apparently sold as a 6.60 ct diamond. The stone had been mounted in a ring but the wearer complained that her finger had erupted after wearing it. Examination showed an SG of 5.272 (diamond is 3.52) and the stone appeared metallic with an RI above the limit of the standard refractometer. With a Geiger counter the specimen recorded 500 counts per minute at a distance of 5cm. When shielded with 0.0012 inch (0.03 mm) of aluminium foil the counts diminished only to 490, showing that γ-radiation was primarily responsible. Exposure to dental X-ray film produced a distinct autoradiograph. X-ray diffraction analysis showed that the material was completely amorphous, probably due to radioactive breakdown, and that the composition was very rich in uranium, resembling that of pitchblende from Great Bear Lake, Canada. The only way to show such a stone is from a lead container, if radiation burns are to be avoided!

A CZ specimen containing inclusions that might have led to misidentification as diamond is reported in the fall 1984 issue of *Gems & Gemology*. The stone, a round 0.90 ct brilliant, *showed irregular swirly growth features somewhat like the graining that may be seen in diamonds. Another CZ seen by GIA contained spherical inclusions oriented in subparallel lines. The inclusions were seen under magnification to be negative crystals but with voids lined, at least partially, with a white material, perhaps undissolved zirconium oxide. In some of the voids, angular growth patterns could be seen.*

In the summer 1988 issue of *Gems & Gemology* the GIA report how they were asked to find whether or not diamond was present in a round yellow metal earring which contained a centre segment of highly reflective tinsel-like flakes embedded in a transparent colourless material, the latter believed to contain diamonds. Examining the tinsel-like particles showed that they were very thin, rectangularly shaped and transparent to transmitted light. An amorphous pattern was obtained for the particles when X-ray diffraction was used so diamond was not the answer. Gas bubbles and an acrid smell when heated showed that the surrounding material was a plastic.

One of the earliest reports on the filling of diamonds is published in the fall 1987 issue of *Gems & Gemology.* Treatment had been detected at the Central Gem Laboratory in Japan, and the filling was understood to be carried out to minimize the effect of cleavages and fractures. In this early report it was believed that silicone may have been used to give a whitish appearance to cleavages and to reduce diffused reflections from them. A diamond received by the GIA in 1987 had been boiled in concentrated sulphuric acid, and this treatment had removed some of that part of a filling which came close to the surface. The 1.22 ct stone displayed a whitish and prominent subsurface cross-like pattern across the table and crown: this probably approached the prefilling appearance of the diamond. In addition, the stone showed an iridescent effect, later to be recognized as one of the best indications of fracture filling.

A diamond crystal of 2.95 ct was reported in the winter 1989 issue of *Gems & Gemology.* The well-shaped bluish-green octahedron was not very clear and did not show the green radiation stains expected on the surface of naturally coloured green diamond crystals. The coloration was superficial and concentrated along the crystal edges. It was presumed that the crystal had been treated by electrons or α-particles. No luminescence was observed under either type of UV radiation: the hand spectroscope gave absorption lines at 594, 504–498 and 415.5 nm, the last indicating that the original colour was representative of a Cape stone.

A report in 1989 highlighted the use of CZ crystals as imitations of diamond crystals. Sales were reportedly taking place in Namibia, and prices up to US $4000–5000 were being asked. Such an imitation would only be saleable in a diamond-producing country, where it is usually illegal for unauthorized persons to own rough diamonds.

Back in 1983 the GIA reported how at a large auction house a 10.88 ct diamond painted with pink nail varnish was substituted for a fine fancy pink stone of 9.58 ct. *If such a fraud is suspected, an acetone-soaked cotton swab will quickly remove the colour, or, of course, nail polish remover, which is more likely to be on hand.* As always, suspicion has to be in the mind from the start.

Evidence of electron treatment in diamond may sometimes show as a zone of colour at the culet or along facet junctions. In the summer 1991 issue of *Gems & Gemology* the GIA cite a yellow diamond of 37.43 ct with a yellow zone in the culet area.

A yellow CZ masquerading as a 10 ct diamond *showed absorption areas close to the 478 and 453 nm bands expected in natural yellow diamond of type Ia. The line at 415.5 nm could not be detected, fortunately and the stone fluoresced orange under LWUV. Set with a diamond in a ring, the difference in dispersion between the two stones could be seen*, as reported in the winter 1990 issue of *Gems & Gemology.*

In a dark-green diamond of 0.50 ct examined by the GIA and reported in the winter 1990 issue of *Gems & Gemology*, cyclotron treatment was indicated *by the characteristic umbrella-like markings surrounding the culet. However, a further coloration could be seen as a zone of darker colour when oblique lighting was used. The darker colour followed the periphery of the stone, extending in a distinct plane from the crown into the pavilion towards the culet. No fluorescence was observed under either type of UV radiation: an absorption line at 592 nm could be detected only when the stone was cooled with a refrigerant. Weak Cape lines could be seen under normal conditions.*

The laser drilling of diamonds has been carried out for at least 25 years, and during that time the pattern of the tracks and the shape of the individual holes has varied widely. In the fall 1990 issue of *Gems & Gemology* the GIA report on a 1.55 ct pale-green marquise which showed natural brown irradiation stains on the girdle. The laser drill hole appeared to have reached an internal cleavage or fracture and branched from that point to give a dendritic pattern. Since natural patterns of this kind have been seen care needs to be taken in examining the hole and track.

While the lasering of diamond is closer to the role of this book than are natural inclusions, one may be taken for the other without difficulty! In the summer 1994 issue of *Gems & Gemology* the GIA report some natural inclusions that were first thought to be laser tracks. *One of the 'tracks' could be seen, under magnification, to be made up of a string of pinpoints which did not break the surface and therefore could not be a laser track. In another example the 'tracks' were shown to have a squarish outline rather than the rounded ones made by a laser.* Some laser holes are not completely straight, and some tracks have been observed to branch, forming a Y-shape.

In October 1991 the General Electric Research and Development Center, Schenectady, New York, announced the synthesis of large gem-quality diamonds composed of 99 per cent carbon-13; natural diamonds are almost entirely carbon-12. The virtually colourless crystals were grown by a combination of chemical vapour deposition and high pressures. Interestingly, the perfection of the crystals was found to be so high that they have great potential for all kinds of industrial and research applications. In turn this means that many crystals will 'spin-off' towards the jewellery trade.

In the winter 1991 issue of *Gems & Gemology* the GIA describe a 0.82 ct dark-green marquise brilliant-cut diamond which showed dark-green natural radiation stains on the girdle at each point. The uniform colour of the stone, which may have been irradiated by neutrons in a reactor, aroused suspicions of treatment. As the stone had natural radiation stains the stone may have been chosen for treatment or it may have begun life with a much paler green colour. Heating following irradiation would have turned the colour to orange–yellow.

Faceted diamonds have been coated with a bluish-grey synthetic diamond thin film to improve their appearance, as reported in the summer 1991 issue of *Gems & Gemology.* Originally colour-graded G and H-1, the stones were coated on the pavilion and part of the crown only, using the hot-filament technique and temperatures of 950°C. Stones became a dark bluish-grey. *With the microscope it could be seen that the coating was uneven on facet junctions, where a whitish appearance resembling minor abrasion was apparent. This cannot be seen on natural blue diamonds.*

Fracture-filled diamonds may suffer damage in the polishing and cleaning process. The fall 1992 issue of *Gems & Gemology* describes the experience of a New York diamond polisher when fashioning a 10 ct rough crystal. Early in the cutting stage the crystal fractured, so the stone was sent for fracture filling. This improved the appearance of the crystal and the polishing process was concluded normally. However, the finished stone showed that some filler had leaked from the fractures, so a second filling was carried out. At this point the stone was recut to a different style, and again some filler was lost. The stone, now weighing 6.90 ct, was filled for a third time. It was then set in a ring and later, when subjected to ultrasonic cleaning and left there by mistake for two hours, a large fracture breaking the table facet could be seen.

Under the microscope it was clear that some of the filler had left the break near the surface on both table and pavilion. Under dark-field illumination the filler itself could be seen to have shattered. Some purplish-pink and greenish-blue flash effects were observed.

We see elsewhere in the book that 'black' (in reality very dark green) diamonds can be treated by substances which leave the stone radioactive, sometimes for very long periods. A stone examined by the GIA and reported in the summer 1992 issue of *Gems & Gemology* was treated with radionuclides in a nuclear reactor. Europium-152, europium-

154 and cobalt-60 were traced: the stone could be worn safely after an interval of about 36.6 years to comply with guidelines set out by the United States Nuclear Regulatory Commission. *Black-appearing diamonds should if possible be tested in a laboratory before purchase.*

The heat from the process of setting a diamond can damage the filler, as a report in the summer 1992 issue of *Gems & Gemology* describes in the case of a 3.02 ct diamond mounted in a ring undergoing repairs. *The stone showed the flash effect (filled stones will show an orange-to-blue or purple-to-green flash when examined in a direction nearly parallel to the fracture).* The heat from the jeweller's torch had in fact made the fractures more visible by damaging the filler.

An unusual method of fracture filling a diamond is reported by the GIA in the spring 1993 issue of *Gems & Gemology*. The 2.51 ct diamond contained a fracture which apparently did not reach the surface and it was at first not clear how the filling had been accomplished. Two small holes were later noticed on the girdle at the 3 and 9 o'clock positions. These had been drilled by a laser, the holes providing access for the filling material. *The stone gave a purple flash, and trapped bubbles could be seen under magnification, effects characteristic of fracture-filled diamonds.*

Black diamonds (green irradiated diamonds with so dark a colour that they appear black) are sometimes found to be radioactive. *Gems & Gemology* for winter 1992 reports that it may be possible to eliminate the radioactivity by prolonged boiling of the stones in acid. This might get rid of metallic polishing residues collected in surface-reaching fractures. Irradiation is said to have taken place in a nuclear reactor: this may have produced radionuclides in the polishing residues.

Rough diamonds as well as polished stones may show evidence of fracture filling. *Diamond Intelligence Briefs* for 24 September 1992 warns that rough diamonds are being clarity enhanced before being shipped to some African diamond-producing countries for marketing. It is rather unfortunate for the perpetrators of this technique that the fillings so far are unable to withstand the heat generated during the polishing process. None the less, buyers of rough should check crystals for the flash effect, trapped bubbles or flow structure, all quite easily seen under magnification.

In the summer 1993 issue of *Gems & Gemology* a green diamond known to have been purchased from a New York City jeweller in the 1930s and to have remained in one family since that time is reported. The diamond was tested for colour origin. *Under the microscope, brown spots and patches could be seen on the surface, especially on the pavilion facets. Green spots and patches commonly result from surface treatment with a radioactive compound – such a process leaves the stone radioactive.* The green diamond gave a maximum reading of 42 mR/h, thus proving that the colour had been artificially induced. Further examination showed that lead-210 was present. When lead-210 is present, the radioactive daughter nuclides bismuth-210 and polonium-210 must also be there. Only lead-210 emits sufficient γ-rays while decaying to be measurable quantitavely. Both lead-210 and bismuth-210 decay by emission of β-particles, polonium-210 decaying by emission of α-particles. Penetration into the surface of a diamond is about 0.01 and 1 mm, respectively. *Surface staining is caused by this penetration and the green colour is found only in a layer at or just below the surface. The brown surface stains shows that this stone had been heated to 550–600°C after irradiation. The ring could have been worn for 357 hours a year without exceeding United States federal recommendations for radiation exposure to the general public.*

A diamond which cannot safely be worn until the year 6507 is described in the spring 1993 issue of *Gems & Gemology*. The stone, a 2.51 ct greenish-yellow round brilliant, had been treated by americium, which has a very long half-life. *The stone showed a distinctive blotchy pattern of lighter and darker areas on the table and gave a residual radiation dose rate of 0.1 mR/h. Under LWUV it fluoresced a strong chalky green but did not phosphoresce. Absorption bands at 595, 504 and 498 nm with the Cape spectrum suggested that the original colour was light yellow.*

A yellow synthetic diamond crystal weighing 0.74 ct was examined by the GIA and showed a predominantly cubic form with minor octahedral and dodecahedral faces. *Gems & Gemology* for fall 1993 reports that *the crystal contained fairly large inclusions with a metallic lustre, and when suspended by a thread was attracted by and adhered to a magnet. Under magnification a square pattern of UV fluorescence and colour zoning could be seen in the middle of the base of the crystal – the fluorescence colour was moderate green under SWUV and a weaker green under LWUV.* With infra-red spectroscopy the diamond was classed as essentially type Ib with some IaA characteristics. Absorptions at 733 and 659 nm with a weak feature in the infra-red are attributed to nickel and are consistent with the crystal having been grown in a nickel-bearing metallic flux. An absorption at 637 nm suggested annealing.

When the GIA tested an 0.88 ct heart-shaped brilliant-cut diamond with low-relief fingerprint inclusions containing tiny voids, it was clear that the stone had been fracture filled. *Under dark-field illumination, as reported in the summer 1993 issue of* Gems & Gemology, *several transparent, colourless filled fractures with the voids could be seen, and an orange-to-blue flash was indicative of a filled stone. However, the flash was hard to see because the fractures were at very shallow angles to the diamond surface. The use of a pin-point fibre-optic light source showed the fractures more clearly and also revealed hairline cracks in the filler. When a polarizing filter was used with transmitted light the outlines of the fractures were easier to see.* Energy-dispersive X-ray fluorescence (EDXRF) – a spectroscopic technique – showed that lead was present, as has been found before in diamond filling material. X-radiography showed up the filled areas as white patches opaque to the rays.

At the Jewelers of America International Jewelry Show held in New York in July 1993, Thomas H. Chatham said that Chatham Created Gems of San Francisco, California, was setting up a plant in Siberia for the manufacture of synthetic gem-quality diamonds. This would be operated by a new company, the Chatham Siberian Gem Company. Polishing of the crystals produced would be carried out either in Russia or in Thailand and stones in a variety of colours and qualities would be available at approximately 10 per cent of the cost of natural diamonds of the same quality.

In the fall 1993 issue of *Gems & Gemology* the GIA report on a selection of yellow synthetic diamond crystals and cut stones produced in Novosibirsk. Discussions with Professor N.V. Sobolev of the Institute of Mineralogy and Petrography, Siberian Branch of the Russian Academy of Science, Novosibirsk, and separately with Mr Thomas H. Chatham, appeared to show that the stones would be grown by a similar technology to that proposed by Mr Chatham.

When a light-green diamond is tested, it is sometimes hard to tell whether the colour (resulting from irradiation) has come about in the course of natural growth or has been induced in the laboratory. In the winter 1993 issue of *Gems & Gemology* the GIA report that *one sign of artificial irradiation appears to be the presence of a small blue zone close to the culet. In a stone of 0.75 ct, an attractive light bluish-green, with a faint woolly absorption near 500 nm (usually associated with brown diamonds) and a weak yellowish-green fluorescence under both LWUV and SWUV, a bluish zone could be seen on turning table-up under diffused light. The bluish zone could be seen only on one side of the pavilion. The stone was in fact a treated light brown and the blue zone was sufficient to give an overall blue–green colour.*

With infra-red spectroscopy the diamond was shown to be a mixture of types Ib and IaA. Two crystals and two cut stones sent to the GIA and reported to have come from Chatham Created Gems Inc. showed characteristics that suggested that they had a common origin with the cut diamond and with known synthetic diamonds from Russia.

In 1993 the World Diamond Congress discussed diamond treatments and their disclosure, the topic being debated by the International Diamond Manufacturers Association (IDMA) and the World Federation of Diamond Bourses (WFDB). A resolution passed by the WFDB stated:

1 The fact that diamonds have been artificially infused with foreign matter, or are coated, or are wholly or partly synthetic, or have been treated by irradiation, must be disclosed as such when offered for sale and in writing on the invoice and memorandum. Any breach of the above rules by a member of an affiliated Bourse shall be regarded as fraudulent.
2 Any violation of the above rule shall be referred to the Bourse for disciplinary action and shall be grounds for suspension, expulsion, fine or such other appropriate disciplinary measure as provided by the by-laws of the Bourse. If the seller alleges that he was not aware of any treatment, he shall bear the burden of proof thereof in order to avoid any sanction.
3 If the seller of a diamond, even in good faith, fails to abide by the above rule, the buyer shall be entitled to cancel the sale, return the diamond, obtain a refund of the purchase price and any direct damage as they, the buyer, may have suffered.

Diamonds treated in a cyclotron are now a matter of history since this process is not now used commercially. However, the treated stones are still around! In the winter 1993 issue of Gems & Gemology *the characteristic features of such treatment are described* (the 'umbrella' effect encircling the culet of a green, cyclotron-treated stone is well-known), and examples are also given of less commonly encountered features. The umbrella effect shows that the treatment had penetrated only a small distance into the stone. *The umbrella is usually placed symmetrically, but one stone in which the effect was asymmetrically placed was ascribed to at least one unit using a beam which did not strike the stone squarely on the culet, so the treater had either to treat the stone a second time after rotating it through 180° or to remain content with only one exposure. The unit concerned, at Columbia University, thus seems to have left its own trade-mark on some diamonds!*

In the winter 1993 issue of *Gems & Gemology* an orange–yellow diamond of 0.34 ct is described. *The stone luminesced a weak to moderate green colour when illuminated by an intense beam of white light. Under LWUV there was a moderate to strong orange fluorescence with zoning and a chalky appearance, the zones being defined by two narrow intersecting cross-like arms of greenish-yellow fluorescence extending diagonally from the girdle edges. Under SWUV the reaction was similar but only slightly stronger. No phosphorescence was observed under either type of UV. The comparative strength of the SWUV reaction compared to that shown under LWUV suggested a synthetic diamond.*

Distinct colour zoning could be seen under the microscope, with darker zones surrounding a central core of lighter yellow and with a funnel shape. The two zones were outlined by a strong dark yellow. When the stone was examined with the table facet down, some near-colourless zones could be seen near the centre and at the corners. A cloud of reflective pinpoint inclusions were present throughout the stone. No surface graining could be seen, but some elongated black inclusions were present, located chiefly at the corners of the stone, as well as one small crystal. Some portions of the original crystal surface showed at two opposite corners, their presence suggesting that the stone had been cut to get the largest possible size. Between crossed polars the diamond showed weak anomalous birefringence, and the stone was attracted by a pocket magnet without adhering to it.

In the summer 1994 issue of *Gems & Gemology* a Russian synthetic diamond is described. The near-colourless stone weighed 0.42 ct, *and showed large metallic inclusions and fluoresced yellow under SWUV. The inclusions were sufficiently large and profuse for the stone to be attracted to a pocket magnet, and there was a persistent phosphorescence for 30–45 seconds after SWUV irradiation.* The stone was identified as type IIa by its mid-infra-red spectrum.

From time to time there have been reports on the instability of some of the materials used for diamond fracture filling. In the spring 1994 issue of *Gems & Gemology* there is an account of a fractured 0.27 ct diamond purchased as a filled stone: *on exposure to a 4 W short-wave UV lamp at a distance of approximately 10 mm a visible degradation, seen as a darkening of colour, was found to be occurring in the filler near to its point of entry into the surface of the diamond. This was detected at a magnification of 40× after 1.5 hours of exposure.* After an exposure of 10.2 hours the degradation could be seen deep in the stone at 10× magnification. After a total exposure of 10.2 hours the discoloration of the filler could be seen with 2.5× magnification.

Diamonds treated by radioactive substances can prove hazardous to those who handle them frequently, and in some countries possession and sale of radioactive substances is made the subject of strict regulations. In the fall 1994 issue of *Gems & Gemology* the GIA describe an 0.43 ct yellowish-green unevenly coloured diamond whose girdle showed two naturals and in which shallow pale-green blotches were visible on the table and on the lower half of the pavilion. These patches were seen most effectively when the stone was immersed in di-iodomethane. The visible absorption spectrum showed a strong GR1 absorption band at 741 nm, this being characteristic of surface-only green coloration induced by irradiation.

The pattern of coloration seen on the facets, however, indicated that treatment with radioactive salts had taken place. The pattern showed characteristic green spotting but no radiation higher than that given off by the room conditions was recorded, as might have been expected. Since the instruments first used for radiation detection might not have recorded all radiation present, the stone was further tested, this time revealing, after three hours of scanning for γ-rays, a small but significant peak from americium-241. Since this is an artificial radionuclide, its presence in the diamond proved that irradiation treatment had been carried out.

Under United States law the diamond may not be sold until the fiftieth century. This is because it is an α-emitter though too low in radioactivity to register on the Geiger counter. Since a diamond may have been treated for colour improvement at any time, *gemmologists and jewellers should not assume that a diamond cut in an old style will not have been treated recently. In this case the abraded facet junctions on the crown of the stone showed that faceting took place before americium was first available, in the late 1940s or early 1950s.*

Gemmologists may expect evidence of treatment only in recently cut specimens, but a note in the spring 1995 issue of *Gems & Gemology* reminds us that all coloured diamonds need to be tested. The note describes an 0.54 ct, old-mine brilliant-cut diamond with a greenish-yellow colour and showing a strong, slightly chalky blue-and-yellow zoned fluorescence under LWUV with moderate to strong yellow fluorescence (with one zone showing slightly blue) under SWUV. No phosphorescence could be seen. These responses were consistent with naturally-coloured diamond but glassy naturals with a melted appearance could be seen on the girdle. Under fibre-optic illumination a yellow graining was seen to be overprinted with a weak green graining .

With the spectroscope the Cape absorption line at 415.5 nm could be clearly seen but other lines usually associated with the Cape spectrum (at 478, 466, 453 and 435 nm) were absent. There was no absorption line at 595 nm but the pair of lines at 503/496 nm could be seen. This absorption pattern strongly suggests treatment but is not entirely conclusive: on the other hand, with infra-red spectroscopy weak but definite H1b and H1c peaks were registered, thus proving irradiation of this type 1a diamond.

For some time the GIA has issued origin of colour certificates for light-yellow and light-brown diamonds, stones with insufficient colour to place them in the fancy colour range. A light-yellow round brilliant weighing 0.91 ct was examined by the GIA and reported in the winter 1994 issue of *Gems & Gemology.* The stone was a moderate blue transmitter and showed a weak green haze without green graining. It showed a moderate localized mottled strain pattern, and fluoresced a very strong blue under LWUV, and a strong yellow under SWUV, with a weak phosphorescence under both types of UV. *The spectroscope gave a moderate Cape spectrum with weak lines at 504 and 498 nm: no absorption was found at 595 nm. This pattern taken with the green haze and lack of graining* led GIA staff to examine the stone further. Mid-infra-red spectroscopy showed that it was a natural stone of type IaB > A with a high nitrogen content. Absorptions at 5163 and 4932 cm^{-1} (the H1b and H1c peaks) proved conclusively that the stone was a type 1a diamond which had been irradiated and annealed. The GIA considered it possible that the stone had undergone treatment to improve the colour without this being satisfactorily achieved.

An instrument called the 'Magnetic Wand' and consisting of a neodymium iron–boron magnet mounted on a wooden shank 60 mm long has recently been reported (in *Gems & Gemology* for spring 1995). The magnet, which is about 5 mm in diameter, has been found to attract the De Beers, Sumitomo and Russian gem-quality synthetic diamonds. The attraction is the result of the use of iron–nickel fluxes in synthetic diamond growth, and *the magnet, devised by W. W. Hanneman, appears to be a useful and simple test for the synthetic material.* The promotional literature suggests that one way of testing might be to make a 'raft' of plastic foam and float the stones in a glass of water. The diamonds can be drawn across the water by the attraction exerted by the magnet. Care should be taken to ensure that the magnet, said to be the most compact magnetic material currently available, does not corrupt magnetic strips on plastic cards placed close to it.

A red natural diamond is extremely rare, and any specimen should be tested for evidence of enhanced coloration. In the spring 1995 issue of *Gems & Gemology* the GIA report a dark-red diamond of 0.14 ct, cut as a round brilliant and *responding to a hand magnet* (this suggesting that the specimen contained large metallic flux inclusions). One such inclusion could be seen under the table facet, and pinpoint inclusions were seen concentrated into wedge-shaped areas. These zones showed a green, hazy appearance, and a similar green zone could be seen in the centre of the stone. Under both LWUV and SWUV the stone, examined through the table, showed a cross-shaped area with moderate green fluorescence, the remainder of the field showing a faint orange under SWUV while remaining inert to LWUV. *The spectroscope showed several absorption lines between 660 and 600 nm with other lines at 635 and 595 nm, and an emission line at about 580 nm. This is characteristic for treated pink to red diamonds.*

One of the stones contained large globules and droplets of residual flux – they were found to contain iron and nickel by the use of EDXRF analysis. Infra-red spectroscopy showed that the stones were a low-nitrogen mixture of type Ib and Ia diamond with A aggregates dominating B aggregates.

The features listed above are consistent with reports on some Russian synthetic diamonds.

Probably one of the first instances of synthetic diamond rough being offered as natural occurred in 1995, when diamond microcrystals were purchased in Canada. The crystals were said to have come from a core drilling in Saskatchewan. On examination the crystals were found to show an elongated octahedral habit with metallic flux inclusions.

Treated pink diamonds have been known since at least 1959, GIA report in *Gems & Gemology* for summer 1995 in a note on a treated pink stone which resembled diamond from the Argyle mine in Western Australia. *The colour of treated pink diamonds is usually highly saturated and specimens show a strong orange response to both LWUV and SWUV. Their absorption spectrum shows sharp absorption lines at 658, 617 and 595 nm with an emission line at 575 nm. This spectrum is diagnostic for treated-colour pink diamond. Some diamonds show uneven colour distribution with distinct zones of yellow and pink, an effect visible with the 10× lens.* To obtain the pink colour, the original stone must be type Ib, containing dispersed nitrogen. These are rare and usually small with a saturated orange–yellow colour before treatment. This colour is usually regarded as valuable in itself, so relatively few stones are offered for treatment. Most synthetic yellow diamonds are also type Ib, and will turn pink if irradiated and heated. An orange fluorescence may also be shown by type IIa pale-pink diamonds: *these may be distinguished from treated pink stones by their lower colour saturation and very faint absorption lines which need a recording spectrophotometer for adequate observation.*

The presence of a plastic fracture filling in a coloured diamond is sometimes hard to spot because the flash colour from the filler may be hidden by the colour of the stone. In *Gems & Gemology* for fall 1995 the GIA report a yellow round brilliant of 1.19 ct in which iridescence was seen when observation was in the direction perpendicular to the fracture plane: the flash effect could only be seen in a direction nearly parallel to the fracture plane. In addition, it was possible to see several colours at the same time in the iridescent area – they were green, greenish blue and purple. The flash colours, on the other hand, could be seen only one at a time, and the colour seen depended upon the background lighting: green was seen in bright-field lighting. Under high magnification the filled part of the fracture showed small bubbles and flow lines but the iridescent unfilled part showed white and feathery with moderate relief. The GIA did not issue a colour origin or colour grade for the stone but stated only that the diamond contained a foreign material in surface-reaching fractures.

Two rectangular modified brilliant-cut and one round brilliant diamond were shown to the GIA in 1995. The report, in the summer 1995 issue of *Gems & Gemology*, described the stones as saturated yellow and with a strong to weak yellow–green fluorescence under LWUV showing as a cross-shaped pattern. A similar but weaker reaction was seen under SWUV, and absorption was observed as vague but increasing towards 400 nm. Two of the stones showed weak green transmission luminescence, and two had weak absorption at 527 nm, seen under low-temperature conditions. All the stones were weakly attracted to a magnet. By diffused light, vague colour zoning of light and dark yellow could be seen under magnification. When all the stones were immersed in di-iodomethane a distinct colourless cross could be seen within a medium-yellow body colour. Among clouds of pinpoint inclusions some were seen to form stringers, an effect not so far reported in natural diamonds.

Early fracture filling of diamond is highlighted in a report in the winter 1995 issue of *Gems & Gemology.* The stone was a modified brilliant weighing 1.07 ct and coloured yellowish orange. The clarity grade was low because one prominent feather crossed the table while others were also present. *Blue and orange flashes were associated with the larger feathers*, and EDXRF spectroscopy determined that thallium as well as the more expected lead and bromine was present in the filler. Thallium had been an early candidate for a filler constituent: while it has been used for highly refractive glasses its toxicity probably prevented prolonged use, and it is less often encountered in current fillings.

The elements iron and nickel are frequently used in at least one method of diamond synthesis, and their presence can be detected in the finished stone by EDXRF. They occur in the metallic inclusions found in many synthetic diamonds, although their presence is not complete proof of artificial origin. The GIA report in *Gems & Gemology* for winter 1995 on a 9.61 ct semi-translucent, marquise-cut black diamond which had been submitted for determination of colour origin. While natural black diamonds are coloured by profuse black inclusions, as this stone was, the inclusions were found to occur in bands which were accompanied by additional near-colourless and brown bands. Brown staining was seen in large fractures.

EDXRF showed that iron was present, perhaps as the cause of the brown stains (perhaps from the polishing process): this is common in polished black diamonds because they are usually heavily fractured. The black inclusions were probably graphite, as no other element was shown by EDXRF (which cannot detect carbon). While the origin of this stone could not have been proved by iron being shown to be present, *the specimen was not attracted to a hand-held magnet, as many synthetic diamonds are*, due to the iron–nickel flux used in their manufacture.

Thin-film diamond coating of gemstones is a possibility, and two techniques are described in the winter 1988 issue of *Gems & Gemology.* Both are experimental, and by 1996 appear to have remained so. One method involves the heating of a metal filament to incandescence in the presence of a mixture of hydrocarbon gas and a hydrocarbon vapour, often methane. The molecules of hydrogen split at the filament into individual atoms, and the hydrocarbon molecules break, allowing some carbon to be freed. The stone to be coated is heated to at least 1000°C. When the hot mix of gases meets the stone, a thin film of carbon, as synthetic diamond, forms over the surface. In the second method the same mixture of gases is used, but involves irradiating them with a radio-frequency field and/or a microwave beam. The need to heat the stone to such high temperatures clearly rules out many gem species being coated in this way. The technique of ion beam enhanced deposition has been used on subjects with low melting points, however.

Coating gemstones with diamond-like carbon (DLC) has been carried out but does not seem to be occurring on a large scale. In the fall 1991 issue of *Gems & Gemology,* DLC is said to be amorphous and with a brownish colour and an RI of about 2.00, too high to register on the normal refractometer. It has a hardness between that of diamond and corundum, and contains hydrogen as a major constituent. Films described in an earlier issue of *Gems & Gemology* are monocrystalline with a bluish-grey colour, an RI of 2.4 and a hardness of 10. These diamond films do not contain hydrogen. DLC films have been applied to citrine, amethyst, beryl, tourmaline, garnet and strontium titanate. The thickness of the film is about 0.08 μm. *When the stone is light coloured, the brownish coating is apparent, and stones with a low RI, such as quartz and amethyst, take on an unusual adamantine lustre.*

Jewellers and gemmologists are always likely to be deceived by very small colourless faceted stones, the kind known as melée in the trade. Particularly dangerous is strontium titanate, whose high dispersion can trick the unwary into imagining it to be diamond, and the same material when used as the pavilion in a colourless composite. Melée comprising stones as small as 0.002 ct has been reported in the literature.

Chapter 8

Ruby and sapphire – the corundum gemstones

The famous gemstones ruby and sapphire are both varieties of the mineral corundum and differ only in their colour. The name 'sapphire' is used for all colours of corundum apart from red; the name 'ruby' means 'red' and the stone has a unique status among gem minerals. There is no difficulty distinguishing between ruby and blue sapphire of course, but all varieties of corundum are routinely synthesized by a simple, quick and cheap process, and since the red and blue colours of natural corundum are of high value in the jewellery world, the ability to distinguish between natural and synthetic is very important. In some respects glass does not play quite so significant a part in corundum imitation as it does in diamond: corundum synthesis is so cheap – stones may cost only a few pence per carat – that manufacturers of cheaper jewellery lines can afford to use the better-coloured and far harder synthetic material.

When we looked at diamond we found that many specimens contained natural mineral inclusions and that a stone purporting to be diamond that did not show similar inclusions should come under suspicion. The corundum gemstones also contain solid inclusions but also characteristic patterns of liquid droplets which can be seen with the 10× lens and which are highly characteristic of the mineral. As with most synthetic gemstones, the chemical composition and physical properties of natural and synthetic ruby and sapphire are the same: for this reason some of the routine gemmological tests do not provide a diagnosis on their own.

Corundum is a hard aluminium oxide with a simple composition which makes it easy to synthesize. Unlike diamond (one of the forms of carbon and the only element to form a gem species), corundum can incorporate different elements which are responsible for the different colours. Manufacturers synthesizing corundum need to incorporate these elements.

For ruby, chromium is needed, and for blue sapphire, iron and titanium combine: iron on its own accounts for the green sapphire and for the yellow colour in natural sapphires. Yellow synthetic sapphires may be coloured by iron or by nickel, sometimes by chromium. Some pink synthetic sapphires are coloured by manganese, but the majority owe their colour to chromium. Vanadium-doped sapphires are offered as 'synthetic alexandrite' or just as 'alexandrite': of all the deceptions practised on inexperienced gemstone buyers, this is by far the most common and particularly affects those with a little knowledge of gemstones, since those completely ignorant of them would not have heard of alexandrite! These products are a completely distinctive slate colour in one direction and purple in the other; far from resembling alexandrite they are more like amethyst.

The range of colours made possible by doping is very large but customer demand limits the number of colours in practice. Without the addition of a dopant, corundum will be colourless and such material could be used as a diamond simulant (not a very convincing one) or, more dangerously, as the crown of a composite stone whose pavilion can then be some softer material with a higher dispersion.

We shall see later that while corundum varieties can be easily and quickly grown, other methods of growth are also used, and these are slow and far from cheap. While a 1 ct ruby made by the flame-fusion process can be grown in the course of a working morning, a stone of the same weight grown by the flux method may take months, and since during that time the melting and particularly the cooling rate has to be strictly controlled (by computer) and the apparatus supervised, the price will reflect all this. The price for a flux-grown 1 ct ruby of good quality could easily reach £100 per carat. And some cost more. This answers the question often asked by students: 'why are expensive growth methods only used for ruby and blue sapphire?'

The answer is that only ruby, of the corundum gems, will really command high prices. Even blue sapphire is not routinely made by any of the expensive growth methods.

Testing with easily operated instruments

Flame-fusion grown (Verneuil) corundum

Since ruby is by far the most important of the corundum gemstones we shall discuss it first, beginning with details of stones grown by the simpler and cheaper method. This is the flame-fusion method, with which the name of Verneuil is permanently associated, although he was not the inventor. Gemmologists are always surprised to find that rubies grown by this method have been around since the last century so there are many specimens in existence. While we have already looked at details of the growth method, stones grown in this way obligingly offer a number of clues to the gem tester – or at least some of them do.

First, remember that *no man-made material will show natural solid inclusions.* While we have to learn recognition of natural mineral inclusions, Verneuil rubies usually show curved growth lines, resembling the grooving on a vinyl record and also curved bands of colour distribution. In addition large, well-rounded and randomly distributed gas bubbles are virtually always present (Figure 8.1). These features should easily be spotted with the 10× lens but, as always, photographs in the textbooks and their accompanying descriptions may lead the student to believe that they are always easy to see. This is not always the case, and when a specimen is hard to identify with the lens, the microscope has to be used. Internal features of ruby are usually easier to see than those in some of the other colours of Verneuil corundum. Students often confuse gas bubbles with crystalline inclusions, but look out for reflection of the microscope lamp on the surface of the bubbles – each bubble will show a bright spot in the same place. Such an effect would never be seen with scattered crystals.

When examining a suspected Verneuil ruby remember to rotate the stone under different types of lighting if you want to see the curved growth lines (it is much easier to see the gas bubbles). Quite often you will not see them simply by looking through the table facet: looking in the plane of the girdle (across into rather than down into the stone) often works better. Similarly, the lines will not be seen in the full glare of the transmitted light in the microscope but will reveal themselves when reflected light is used. This can be obtained either by dark-field illumination or by using a free-standing fibre-optic source which can be moved about to get the best effect.

All coloured crystalline substances possess the property of pleochroism, which means a difference in colour seen when the direction of viewing alters. In ruby the two pleochroic

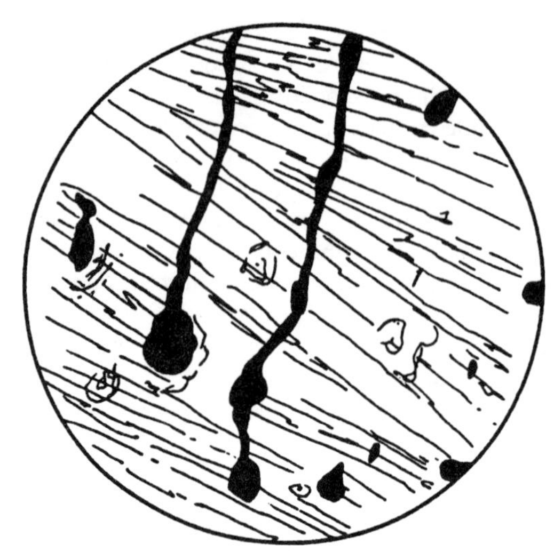

Figure 8.1 Characteristic tadpole-shaped gas bubbles in a ruby grown by the flame-fusion process

colours are a crimson and an orange–red. While the eye can sometimes detect them when the stone is moved under a source of light, it is easier to use the dichroscope, a small instrument whose operation is more fully described elsewhere (see Chapter 5) but which shows the two colours side by side. The dichroscope is very useful in the detection of Verneuil ruby since, because of the way in which the faceted stone is cut from the grown crystal, the two colours will be seen when the stone is examined in a direction at right angles to the table facet. With most (not all) natural rubies, the shape of the original rough crystal forces the lapidary to facet the stone with the table facet in a direction at right angles to that seen with the Verneuil synthetic ruby. This direction happens to be one in which, due to the crystal structure of corundum, pleochroism cannot be seen.

While we have discussed ruby first, the same features apply also to blue and green sapphire and to some of the other, darker colours. With colourless and yellow Verneuil stones the curved growth lines are very hard to see and may even be invisible when sought with the 10× lens alone. While there are microscope techniques which will aid your search, remember first that natural yellow and colourless sapphires will contain natural solid inclusions. One technique is simply to examine the stone with the lens against a white background – a sheet of white paper will do. With yellow sapphires a blue filter placed between the light source and the specimen will sometimes persuade the curved growth lines to appear: the stone needs to be immersed in a clear liquid whose refractive index (RI) is as close as possible to that of the sapphire. Deep blue frosted glass filters are reported to give the best results. Even a pen-torch held a metre or so above the specimen which is placed on a photographic film may produce the lines on the resulting photograph.

The technique of immersion contrast, in which the specimen is immersed in a liquid of known refractive index (di-iodomethane is useful since its RI is close to that of corundum for blue and violet light) and then lit by a parallel and narrow beam of light, is useful in showing up curved growth lines. The light passes through the liquid and specimen to fall upon a photographic film placed beneath. Anderson in *Gem Testing* (10th ed., Butterworth-Heinemann) recommends the use of the light from a photographic enlarger with the stop set down to *f*22. Even the lines in colourless sapphire have been revealed by this method. Straight lines which intersect with the curved lines in some colourless synthetic corundum help in identification although their origin is uncertain. When testing suspected synthetic corundum by these methods the specimen should be examined in different directions. This all takes time, and if the stone is seen to be virtually inclusion-free there is a strong supposition that it is synthetic.

Using the polariscope, a simple instrument with two pieces of Polaroid superimposed with space for the specimen in between, an interesting effect is sometimes seen. The Polaroids need to be set with their vibration directions at right angles to each other (see Chapter 5) so that no light passes through them to the observer.

The specimen is placed between the two Polaroids and the instrument lamp switched on. Once the specimen is correctly positioned, systems of straight lines can be seen inside the specimen as it is rotated. The lines of the second and third system are seen to be at 60 or 120° to those of the first system. Even with colourless corundum, fine lines can sometimes be seen, but they quickly vanish when the stone is moved. In this case a sophisticated microscope incorporating the two Polaroids needs to be used, and stopping down the iris diaphragm allows the lines to be seen in the colourless material.

The hard part of setting up this test is the positioning of the specimen. The optic axis (direction of single refraction) needs to be found first and placed at right angles to the two Polaroids. The optic axis will be found when a group of spectrum colours is seen: when the specimen is placed in such a way that the coloured area is on the top, magnification will show a black cross with coloured concentric circles at the centre. Then the groups of lines can be investigated. The whole phenomenon is known as the Plato effect. So far it has not been reported in natural ruby.

Corundum made by other methods

While the Verneuil flame-fusion method of ruby and sapphire growth accounts for over 90 per cent of total synthetic corundum production, other methods are used to grow ruby (hardly ever blue sapphire and even more rarely the other sapphire colours). Without detailing the production methods here – they are described in Chapter 4 – we should remember that growth takes months rather than hours and that the crystals produced are expensive, reflecting the time and cost of research and manufacture.

Since ruby is by far the most important of the corundum gemstones, the description following should be taken to mean ruby rather than any of the sapphires.

Flux growth

By far the majority of the higher-quality synthetic ruby is grown by the flux method (or flux-melt method). We have seen how this is done elsewhere in the book but for the purpose of simple identification we have first to compare the properties of this product with those of natural and flame-fusion ruby.

Again, the flux-grown stones do not show natural solid inclusions: there are no crystals of spinel, calcite, corundum itself or rutile. Flux-grown rubies show no curved growth lines or curved colour banding as the Verneuil stones do, and pleochroism seen through the table facet is not necessarily an indication of synthesis. This is because the flux-grown crystals take on a wide variety of shapes unlike the Verneuil crystals which are always in the boule form, from which faceted stones all take the same orientation. Plato lines cannot be seen in flux-grown rubies.

When rubies are grown by the flux method, traces of the flux are almost invariably present in the finished stones and appear as metallic structures, often forming twisted veil- or smoke-like patterns. While these can sometimes resemble the flat planes of liquid droplets seen in many natural rubies (these are familiarly known as 'fingerprints' or 'feathers'), moving the stone under a single light source (that provided by fibre-optic transmission is very well suited for this type of examination) will show the opaque and metallic nature of the flux particles.

Depending upon the quality of the rubies, traces of the inclusions can be profuse or sparse: in high-quality material the few flux inclusions, sometimes resembling scattered breadcrumbs or paint-splashes, can be placed at the sides of the faceted stone, so that through the table the stone appears inclusion-free. Such an appearance would be rare in a natural ruby, and a clean stone should always be suspected. Very profuse flux inclusions, ironically, are easier to see, and with practice the gemmologist can confidently identify their nature.

The nature of this particular growth process, which involves a platinum or iridium crucible, often leads to small fragments of the crucible material becoming incorporated into the growing crystal. These appear metallic by reflected light, and are opaque and black in other types of illumination. They are often angular and cannot easily be confused with any natural inclusion.

Both the metallic fragments and the particles of flux are quite easily seen and recognized by the gemmologist using the 10× lens: small stones, as always, need the microscope for diagnosis to be certain.

One of the obvious considerations to be remembered when a synthetic product is examined is that the crystal grower can control the composition of the specimens, and if iron, for example, is to be excluded, then this can be done. In practice, ruby crystal growers do try to exclude iron since, with an appreciable iron content, the unique ruby red is less attractive. Chromium is the cause of colour in ruby, and the use of the hand spectroscope (direct-vision spectroscope) quickly shows whether or not a red stone is a ruby rather than red spinel, garnet, tourmaline or glass. This is all very well, but the spectroscope will not show whether the stone is natural or synthetic! Practically, this may be thought to limit the use of this handy instrument, but this not quite the whole story. The absorption spectrum of ruby is distinctive, and one of the first to become familiar to gemmology students. If the spectrum is very sharp and clear, with no blurring of the absorption bands, there is at least a suspicion that the specimen may be synthetic, and this can then be quickly checked with a lens or microscope.

The use of ultra-violet (UV) radiation was described in the chapter on diamond, and it is also useful for the testing of ruby. When a natural ruby is chrome-rich and iron-poor it will not only provide a sharp absorption spectrum but also glow a strong red under long-wave UV (LWUV). Iron not only causes the colour of ruby to tend towards brown but when present in appreciable amounts 'poisons' luminescence: when the ruby is very iron-rich it will scarcely respond to UV and will be virtually inert. Synthetic rubies, rich in chromium and with iron successfully excluded, will glow very strongly. Even this effect is not quite diagnostic since there is always the possibility of encountering a chrome-rich, iron-poor natural stone with no apparent natural inclusions! Fortunately, after irradiation with X-rays, a synthetic ruby will phosphoresce strongly after the rays are switched off. Iron-poor stones are also much more transparent to UV.

So far then, the simpler gemmological tests, combined with an approach of suspicion to any very clear (and especially any large clear) ruby, will solve the natural/synthetic problem. Ruby overall is not the hardest gemstone to test. Even so, flux-grown stones are being produced and sold at high prices: at least one grower is producing beautiful crystals whose faces give a very natural appearance: since so few ever see gem-quality ruby crystals (they are usually fashioned at or near the source) any fine red crystal with a profusion of faces is likely, through ignorance, to be taken as natural. The crystals are described below. Some growers have produced clusters of ruby crystals but these usually show a bladed habit which would of course make them impossible to facet.

Early manufacturers of flux-grown ruby were Carroll Chatham of San Francisco, California, and Ardon Associates of Dallas, Texas. Both firms were selling their products in faceted form by the mid-1960s under the names Chatham Created Ruby and Kashan Synthetic Ruby. While the circumstances of both firms have changed (Thomas Chatham has succeeded his late father) the products are still appearing on the market.

Writing in 1980, Nassau, in *Gems Made by Man*, said that the flux-grown ruby commanded 100 times the price of the Verneuil product and that comparatively small amounts are produced.

The method of hydrothermal growth in which the desired material is grown in the presence of water in a sealed pressure vessel (the technique is described elsewhere) has not been used commercially for the production of ruby, though some experimental crystals are in large collections. As in the flux-growth process, iron needs to be excluded, so the pressure vessel (autoclave) is lined with silver or platinum: also as in the flux method, growth takes place on a preshaped seed, but if the growth conditions are not ideal the crystals will show veiling and cracking, unlike anything seen in natural ruby. As always, natural inclusions will be absent.

Far more commonly used for the growth of ruby and with a product so clear that any gemmologist ought to be suspicious, is growth by crystal pulling, often known as Czochralski growth. Here a ruby seed is lowered to the surface of a melt and then slowly drawn upwards, taking the melt with it to form a rod or cylinder which is primarily used in the ruby laser. Such crystals contain no flux inclusions since no flux is used, and the only possible indication of this method may be a few elongated gas bubbles. Very fine grooving (striations) may be seen, with some difficulty, under magnification. Pulled crystals are so clear that they have to be artificial – no mineral inclusions of course – but any ruby of this clarity should arouse suspicion.

All the products above are characterized by the absence of iron, of natural inclusions and often by the presence of features hard to see and to interpret. For this reason, although natural rubies almost invariably look ‘busy’ inside, we shall examine some well-known examples more closely.

Knischka rubies

Professor Paul Otto Knischka of the University of Steyr, Austria, began to grow ruby crystals in the early 1980s, using a growth method which he admits is from the melt. Growth produces attractive crystals with numerous faces, a feature not found with any other type of corundum growth. The faces have a notably bright lustre which is rarely if ever found in natural ruby: this is worth bearing in mind since very bright gem-quality ruby crystals would be exceptions on a market which in any case scarcely exists away from the place of origin. Details of the crystals and their faces can be found in a paper by Gübelin in the fall 1982 issue of *Gems & Gemology*.

Knischka ruby crystals have been cut, however, and gemmologists will want to know how the fashioned stones can be identified. The earlier Knischka rubies at least had a violet tinge with the red: this is not unknown in natural ruby but is certainly rare in other synthetic products. While the stones show marked pleochroism, this again is not enough to distinguish them from natural rubies. Under UV radiation the stones gave a strong, clear carmine red, while there was a marked phosphorescence after subjection to X-rays: this is characteristic of iron-poor manufactured ruby. The specific gravity in the range 3.971–3.981, with an average of 3.986, and the RI in the range 1.760–1.761, 1.768–1.769 with a birefringence of 0.008, provide no distinction from the natural material.

Examination of the interior of the rubies gives the only clue to their artificial origin. Liquid 'feathers', colour swirls, negative crystals and black metallic platelets, with two-phase inclusions, together provide sufficient means of distinction from natural ruby while not making an absolute diagnosis of the Knischka product from Kashan, Chatham or Ramaura stones.

Negative crystals (hollow tubes of crystalline shape) perch alone or in groups on the ends of long crystalline tubes and these may become known as characteristics of the Knischka ruby. While early specimens showed metallic hexagonal platinum platelets, Professor Knischka stated in 1982 that he would be able to prevent them occupying crystals in due course. Large gas bubbles which are easily visible under low magnification show that they are the gaseous part of two-phase inclusions, when examined under higher powers. The two-phase inclusions are hard to find inside the stones because their outlines are very fine, indicating that their composition must be a highly refractive material, and the inclusions so far seem especially characteristic of the Knischka ruby.

In summary, this product resembles the Myanmar (Burma) ruby quite closely but contains no natural mineral inclusions. Metallic black platelets and two-phase inclusions with prominent gas bubbles, and long tubes ending in negative crystals all serve to indicate the Knischka stones. Bright and multifaced crystals with no attached matrix (rock) should be viewed with great suspicion if offered as gem-quality natural ruby.

Ramaura rubies

During the early 1980s, fine ruby crystals were manufactured by the Ramaura Division of Overland Gems, Inc., of Los Angeles (later the J.O. Crystal Company). The stones were marketed under the name Ramaura from the start, the company stating that they intended to market faceted stones along with lower-quality material for fashioning into cabochons, single crystals and crystal clusters. In the first major report on this product, appearing in the fall 1983 issue of *Gems & Gemology*, the process of manufacture was stated to involve high-temperature flux growth with spontaneous nucleation. For the purpose of the 1983 paper, 160 faceted Ramaura rubies were examined in the weight range 0.15–7.98 ct and 82 Ramaura crystals weighing from 0.21 to 86.73 ct.

While most gem-quality rubies (though not the Verneuil type) are grown on prefashioned seeds, the Ramaura product is said not to involve growth on a seed or on seed plates. There are advantages in this method of growth when compared to growth by spontaneous nucleation where the crystal grows 'on its own'. The use of a seed allows greater control of the growth rate and of the perfection of the final crystal. While growth by spontaneous nucleation may give rise to different crystal habits (preferred shapes), crystals grown in this way often contain fewer inclusions than seed-based ones. Spontaneous nucleation can arise from crucible wall irregularities or even from specks of dust, and is not always to be discouraged. Crystal clusters can grow in this way, and have enterprisingly been sold on their own in jewellery.

Ramaura rubies range in colour from a pure red through orange–red to a slightly purplish red. Some faceted stones have shown areas of a lighter red, rendering the whole specimen pink. On grounds of colour, there is no way in which a Ramaura ruby differs from natural stones. Inclusions are less prominent, probably because no seed is used: with a seed some element of forced growth is necessary, and this is more likely to trap impurities. Some Ramaura stones appear inclusion-free while others contain visible traces of flux. Tilting the stone shows inclusions and colour zoning most effectively.

While Verneuil rubies are almost invariably cut with their table facets parallel to the optic axis of the crystal (this means that dichroism can be seen through the table), Ramaura rubies appear to be cut randomly, spontaneous nucleation giving crystals with different orientations.

The RI of the Ramaura rubies was 1.762–1.770 and 1.760–1.768 for two stones tested by the Gemological Institute of America (GIA). The birefringence was measured as 0.008. All these figures could represent natural ruby. The SG was also normal for all types of ruby, at 3.96–4.00. Under LWUV the Ramaura rubies showed a variable intensity fluorescence, ranging in colour from moderate to extremely strong dull chalky red to orange-red with some small zones showing chalky yellow. There was no observable phosphorescence. Under short-wave UV the fluorescent colours are more or less the same as under LW, with additional chalky, slightly bluish-white zones in a few specimens. No phosphorescence was observed.

Under X-irradiation some areas were unresponsive and there was no phosphorescence (Verneuil stones usually show a noticeable phosphorescence on subjection to X-rays). While the colours observed are similar to those seen under UV, they are not strong. Providing the chalky yellow and bluish-white zones are not polished away, they provide one of the more characteristic signs of Ramaura ruby.

The absorption spectrum gives no clue to artificial origin, and the stones do not transmit SWUV radiation characteristically enough for this test to distinguish them from natural rubies, although most synthetic ruby does transmit SWUV more effectively than most natural specimens.

As usual with synthetic stones we are forced back upon a study of the inclusions. Here the Ramaura ruby does not let us down! Black metallic flakes from the crucible wall are not present as in the Knischka and many other flux-grown rubies: on the other hand, large inclusions of flux can be seen. Some of them show an orange–yellow colour while others are colourless: there is a great range of size, some flux inclusions being mere particles and others large and drop shaped. Some flux-filled negative crystals may be present, either angular or rounded in shape. Some of the coloured flux inclusions may contain whitish areas which may be near-transparent or opaque, and some have a crackled appearance: this effect is best seen in channels or voids which are partly filled with flux material.

Some flux inclusions may superficially resemble the ‘fingerprints’ (flat features consisting of liquid droplets) seen in much natural corundum. Flux particles forming these features are white rather than yellow. Resemblance to two-phase inclusions is superficial, as close examination shows such features to be completely solid. White wispy veils of flux are characteristic of at least the earlier Ramaura stones.

A very varied picture of growth patterns characterizes many Ramaura rubies. Students beginning a study of a suspected stone should make as much use as possible of the lighting conditions available, changing from dark field to direct transmitted light and moving the (preferably fibre-optic) light source about. Signs of twinning and parting show as parallel lines which may take an angular arrangement, but other shapes have been recorded. A stone from which flux-included sections have been cut away will almost always show some form of colour zoning. Again, the specimen must be examined from every angle as some of the phenomena are very elusive. Since Ramaura rubies can be large and attractive, distinction from natural stones and from other synthetic rubies is especially important, though, as always, natural rubies will show natural mineral inclusions. Those familiar with the polysynthetic (parallel, page-like) twinning lines seen in some natural ruby will find that they appear to penetrate the stone and so can be seen from many angles: conversely, the lines of the colour zoning seen in the Ramaura ruby appear and disappear with change of focus. Such an evanescent effect is also seen with the curved growth lines in Verneuil stones and also features in the Kashan rubies described later.

A combination of single straight growth planes extending into the specimen and which are combined with curved or irregularly shaped growth features is a characteristic of some Ramaura rubies. Tiny particles of undissolved flux forming 'comet tails' (GIA description) have been seen in both Ramaura and Kashan rubies as well as in some natural stones. In the latter, the 'tails' often appear to follow an included crystal whereas in the synthetic rubies they have no 'quarry'.

The Ramaura ruby is not in general difficult to distinguish from the natural stone but an apparently clear ruby will always pose problems, even though inclusion-free natural ruby is rarely encountered. Careful examination with the microscope is the key to identification.

At the time of first release on to the market, the company said that a constituent would be added to the crystals which, on examination, would disclose their artificial origin. So far this does not seem to have happened. The idea was that a rare earth element would be added to give a yellowish-orange fluorescence. Such an addition would almost certainly have given a rare earth absorption spectrum but none has yet been reported. While such a material, if applied as a coating, could have been polished away, it would still be present on unfashioned crystals or clusters.

Kashan rubies

In the mid-1960s, F. Truehart Brown of Ardon Associates Inc., of Dallas, Texas, made the first Kashan rubies. Crystals, crystal groups and faceted stones appeared on the market, and over the years various claims were made for them – one being that they were indistinguishable from natural rubies. To support this claim a very few sets of comparison stones were put together. In 1984, bankruptcy proceedings stopped production, but recently a take-over has allowed manufacture to recommence. When the Kashan rubies were first examined they were found to show clear signs of flux-melt growth, with flux particles forming 'paint splash' inclusions; smaller, dust-like flux particles resemble breadcrumbs and sometimes go by that name. The paint splash inclusions, in early products at least, had an overall moccasin shape and were arranged in parallel groups. Stones were found in the Bangkok gem markets back in the 1970s.

Many Kashan rubies show highly characteristic veiling or an effect resembling smoke in a still room: both effects are caused by particles of flux, and can be seen in many flux-grown gemstones, emeralds as well as rubies. When suspected, such effects should be carefully examined under magnification to ensure that the veils or smoke clouds are twisted and do not form flat planes. Twisting virtually confirms a flux-growth specimen, but the absence of natural mineral inclusions is, as always, an important clue. Another effect, resembling heat shimmer from a warm surface, can be confused with the twinning lines seen in some natural rubies. Strong dichroism is noticeable, with one of the colours a strong orange or brown. These two colours sometimes concentrate in different parts of the stone. While many synthetic rubies are transparent to SWUV, at least one report suggests that some Kashan stones, perhaps with added iron, do not show this effect. The included iron theory is given some credence by the presence of a recognizable iron absorption spectrum in some specimens.

Douros rubies

One of the latest synthetic rubies to appear on the market was reported in 1994. The Douros synthetic ruby is manufactured in Greece by the brothers John and Angelos Douros, who have a background in refining precious metals. The rubies were first shown in 1993.

The growth method is reported to employ two furnaces in which growth proceeds very slowly with slow cooling. Various elements are used to dope the crystals in order to achieve colour effects as near as possible to natural rubies. The method is said to achieve crystals up to 20–50 ct, with the largest recorded crystal (1994) at 350 ct and the largest faceted stone at 8.5 ct.

Crystals take two main forms, some being tabular and others rhombohedral. The latter are reasonably equidimensional and are more suitable for cutting. These habits closely approach those found in the Ramaura product, and a penetration twinning seen in the Douros ruby is also found in Ramaura crystals. Other ruby crystals do not show these habits, however.

The Douros rubies range in colour from a saturated red to purplish red and reddish purple, as reported in the summer 1994 issue of *Gems & Gemology*. Addition of different colouring elements leads to colour distribution in zones with geometrical boundaries. Growth sectors observed in the tabular crystals consisted of near-colourless to light-red outermost zones confined to certain faces, with near-colourless to light-red triangular growth sectors confined to twin boundaries; and purple to bluish-purple colour bands parallel to both the rhombohedral faces. In both crystal types, purple to bluish-purple intersecting acute-angled triangles could be seen, the purple to bluish-purple bands and triangular zones indicating a blue sapphire component within the ruby.

The SG of the Douros ruby falls in the normal ruby range, with measurements of 3.993–4.029. Similarly, the RI covers the range 1.760–1.774 with a normal average birefringence. Some variations in RI have been observed in different portions of the crystals. Faceted stones give an intense orange–red fluorescence under LWUV, with a moderate red seen under SWUV. In some portions of the crystals the top layer is inert, this effect occurring under both types of UV radiation. In practice the inert outer areas of the crystals are polished away when faceting takes place so that the finished stones are usually a uniform red under UV.

Under magnification the Douros ruby shows internal growth planes which parallel the larger crystal faces. Similar features can be seen in Chatham and Ramaura rubies, but the Douros material shows some individual features. In the rhomobohedral crystals areas of different colours with sharply defined boundaries can be seen, the colours varying from a very deep red to a near-colourless to light red. A ruby with such variations of colour so sharply defined strongly suggests a Douros product. The rhombohedral crystals also contain acute-angled purple to bluish-purple triangles lying in the red core. These areas show that a blue sapphire component is present, the colour being caused by interacting titanium and iron.

In the tabular crystals both regular and irregular colour distributions have been observed. There was no colour zoning in the cores of the crystals. Purple to bluish-purple acute-angled triangles are found in some of the tabular crystals.

Some Douros rubies appear to be inclusion-free while others contain residual flux particles either as large pieces or as veils of droplets. The large flux inclusions have been shown to be rounded to elongated cavities filled with a yellowish material containing bubbles or voids. The larger yellowish flux residues show a crazed or mosaic patterning. Overall, the GIA found that the internal features of the Douros stones resembled heat- or borax-treated natural ruby. No platinum crucible material was found in any of the samples.

The spectroscope gave a normal ruby spectrum and therefore will not distinguish Douros rubies from natural stones. Beyond the testing ability of the gemmologist but interesting to the scientist is the GIA's finding that lead nitrate was present in some of the yellow flux material. It would have been formed when nitric acid was used to separate flux from crystals at the end of the growth run.

The Douros ruby should present no serious problem to the gemmologist: the residual flux inclusions are usually present, even though they are not always large and prominent. Most of the flux assemblages contain gas bubbles. Internal umbrella-shaped growth zones are characteristic, and the crystals show faces not seen on natural ruby. No trace of rutile needles has yet been recorded.

Lechleitner rubies and sapphires

For many years Johann Lechleitner of Innsbruck, Austria, has been producing synthetic emerald overgrowth on beryl seeds – and a variety of other interesting crystals, including 'complete' emeralds. In the mid-1980s, Lechleitner grew complete rubies and blue sapphires which do not seem to have entered the market in large quantities; crystals have also been sold. Lechleitner told the GIA in a letter of 1985 that he had been growing synthetic ruby and blue sapphire since 1983 and that all his material had been sent to Professor Dr Hermann Bank of Gebrüder Bank, Idar-Oberstein, Germany.

Lechleitner is also reported to have grown colourless corundum and 'padparadschah' (pink with some orange), yellow, green, pink and 'alexandrite' colours. Some stones have been sold in Japan.

Examined by GIA and reported in the Spring 1985 issue of *Gems & Gemology*, the ruby, weighing 0.47 ct and cut as a round brilliant, showed a strongly saturated purplish red and was transparent with some haziness. Under magnification, flux inclusions could be seen although the stone appeared clear to the unaided eye. The optic axis was nearly parallel to the table facet, so that dichroism could be seen in this direction. The SG at 4.00 is slightly higher than that of some synthetic rubies but not out of the possible ruby range. The RI was 1.760–1.768 with a birefringence of 0.008, also quite normal for ruby. Under LWUV the ruby gave a strong red colour with no phosphorescence: under SWUV the stone showed a moderate red fluorescence with a slightly chalky white overtone and no phosphorescence. No phosphorescence was observed after subjection to X-rays. The absorption spectrum was also normal for all types of ruby.

As always, it is the presence of flux inclusions that marks the Lechleitner ruby as a synthetic product. Wispy veils and fingerprint flux traces were quite easily detected under magnification. Curved growth striae were also seen, a feature associated principally with Verneuil flame-fusion corundum.

The GIA speculated on whether a small Verneuil-grown seed crystal could have initiated the flux growth. This would account for the curved striae if the seed remained inside the finished stone. Alternatively, a larger Verneuil-grown crystal, colourless or chromium doped to give the ruby colour, could have had flux-grown ruby deposited upon it. It is possible that inclusions characteristic of the flux growth process might have entered the Verneuil material.

Such experiments have been made: in the late 1960s a Verneuil–flux ruby was presented to the Natural History Museum in London, and Chatham has also made similar specimens. Among other materials produced by Lechleitner are overgrowth of synthetic pink corundum on Verneuil colourless corundum, synthetic ruby over Verneuil ruby and synthetic ruby over natural corundum. The stress cracks resembling crazy paving, traditionally associated with the Lechleitner emeralds on beryl, have not been observed in either the ruby or the blue sapphire of this study.

The blue sapphire examined was a round modified brilliant of 0.69 ct. The colour was a strong violet–blue parallel to the *c*-axis, which was about 20–30° from the table facet plane. A pale greenish-grey–blue was observed at right angles to the *c*-axis. The sapphire was inert to LWUV and no phosphorescence was seen. Under SWUV a very weak chalky whitish-blue fluorescence occurred with no phosphorescence; this effect was also seen when the sapphire was placed under X-irradiation.

Examination with the spectroscope showed no absorption bands though a broad absorption region was seen to cover some of the far red and some of the violet sections of the spectrum. The SG was 4.00 and the RI 1.760–1.768 with a birefringence of 0.008.

Inside the stone the pattern was similar to that seen in the Lechleitner ruby, but with a greater profusion of flux. Curved colour banding could be seen.

A blue sapphire with no absorption in the visible portion of the spectrum could be a Verneuil product, but the inclusions prove the Lechleitner sapphire to be flux grown. The sapphire may perhaps deceive more than the ruby, as it has a crowded interior.

Chatham blue and orange sapphires

Commercially there is not a great deal of point in growing corundum varieties other than ruby by expensive methods. None the less, from time to time coloured sapphires have been grown. In the 1970s, Chatham grew blue sapphires with the flux-melt method, and though very few stones are on the market, it is worth remembering that they exist. A faceted stone that I examined some years ago showed profuse flux inclusions which, perhaps by chance, appeared in a near-hexagonal pattern. Chatham also synthesized orange sapphire and produced both faceted stones and crystal groups.

The faceted blue sapphires show marked colour zoning with colours ranging from near colourless to light blue to very dark blue. Many areas show whitish from inclusions. In the crystal groups there is also considerable colour variation, with some individual crystals showing no colour while others are dark: some crystals are transparent and others opaque. The crystal groups are coated on the back with a transparent glassy substance in which gas bubbles can be seen. Chatham says that a liquid silica-based ceramic glaze is applied to the back of the groups, the crystals then being fired at 1000°C. The glaze is not applied to single crystals.

The orange faceted stones show pleochroism with strong pink–orange and brownish-yellow colours. Under LWUV a variable orange response is observed, ranging from strong to very strong. Some zones show a chalky yellow fluorescence. Similar colours, though much less strong, can be seen under SWUV. While there is no phosphorescence under X-rays the colour response again varies, with reddish-orange areas and with some portions inert. The absorption spectrum shows the presence of chromium but is not diagnostic for this material: natural orange sapphires are not particularly common but they may show a similar absorption spectrum to the Chatham stones, including for some features that can be ascribed to iron, which is also present in the synthetic stones.

Some of the crystal groups showed a slightly lower value for SG than would be expected from corundum alone. This is due to the ceramic glaze, which the GIA found to have an SG of 3.08. This applies to both the orange and blue sapphire groups.

The flux inclusions are clear signs of flux-melt growth: the blue sapphires contain whitish needle-like inclusions and the hexagonal patterns already mentioned. In both blue and orange stones such characteristic inclusions as colour zoning, platinum fragments, healed fractures and whitish clouds have been noted. The thin whitish needles were seen only in the blue sapphires. Immersion in a suitable transparent liquid reveals the near-hexagonal colour zoning particularly well.

Star rubies and sapphires

The Verneuil process, by which the bulk of synthetic corundum crystals is made, also produces star rubies and blue sapphires. For this, titanium needs to be added to the starting material since the needle-like crystals which form the star are the mineral rutile, titanium dioxide (TiO_2). The rutile needles are arranged in three groups at 120° to each other, and will scatter light when present in sufficient numbers. The star effect is seen when the stone, cut as a cabochon, is examined under a single point of light (not in diffused lighting). Most star gemstones show the star by reflected light (the effect known as epiasterism) while a few others, such as star quartz, show the star best by transmitted light (diasterism). Most stars in corundum have 6 rays, though 12 can sometimes be detected.

In synthetic star corundum the star is unusually sharp and well centred, and shows up strongly against a suspiciously bright body colour. In many natural star rubies and sapphires the body colour is dull and not the finest red or blue. A natural star corundum with fine translucency, a well-shaped and centred star and fine colour is one of the most expensive gemstones on the market, and its synthesis is not surprising. When a Verneuil-made star corundum is examined – its attractive appearance arousing suspicion, we hope – it will show the typical features of large randomly placed gas bubbles, curved distribution of colour and curved growth lines, should the stone be transparent enough.

Linde stars

In 1947 the Linde Division of Union Carbide Corporation began to produce star ruby. Reports at the time said that the best star stones were produced when the rutile was from 0.1 to 0.3 per cent of the total feed powder and when the grown crystals were then kept at a temperature of 1100–1500°C for several hours, in order to allow the needles to crystallize out. This is always the way in which star stones are grown, with the ruby crystal, in which the rutile is disseminated, grown first and then reheated to allow the rutile crystals to form the stars. The finished cabochon should show the star centrally in the exact centre of the dome, and this is ensured by regular temperature variations. Before 1952 the Linde stars were much more transparent than their later products and were thus rather more like good-quality natural star rubies. Stars of the pre-1952 period were rather vague, and curved growth lines were prominent in the

Synthetic, Imitation and Treated Gemstones

Michael O'Donoghue

Butterworth-Heinemann
Linacre House, Jordan Hill, Oxford OX2 8DP
A division of Reed Educational and Professional Publishing Ltd

 A member of the Reed Elsevier plc group

OXFORD BOSTON JOHANNESBURG
MELBOURNE NEW DELHI SINGAPORE

First published 1997

British Library Cataloguing in Publication Data
A catalogue record for this book is available from the British Library

ISBN 0 7506 3173 2

Library of Congress Cataloguing in Publication Data
A catalogue record for this book is available from the Library of Congress

Composition by Genesis Typesetting, Rochester
Printed and bound in Great Britain

stones. Rare star stones in purple, green, pink, yellow and brown colours were produced by Linde but they can never have been common.

Kyocera stars – Inamori stars

Since star stones are not as popular in some markets as in others, the impetus for growers to manufacture many different types is not so strong, so a new product is of particular interest. In 1988 the Japanese firm of Kyocera began to grow star rubies which were marketed under the trade name of Inamori. The stones show quite intense stars of a whitish colour, and the stones also contain imperfections which might lead the unwary into mistaking them for natural specimens. Some stones have had slightly pitted surfaces and sometimes the rays of the star appear to be broken or wavy. The body colour is a slightly purplish red, again similar to natural ruby. A first look at the stones shows their flat backs which are semipolished. Most natural star stones have rough and convex backs, left that way to increase weight. The manufacturer would have only to roughen the backs should the desire to imitate the natural stone be the over-riding consideration. It is probable, however, that the aim is to make a satisfactory star stone and not to attempt serious deception.

Under LWUV the Kyocera star rubies show a very strong red fluorescence, and a very strong red response under SWUV, with a moderate to strong chalky blue–white overtone. Inside the stones very fine rutile needles can be seen: they are notably finer than similar needles expected in natural star rubies. Round and distorted gas bubbles can also be found. Examined with fibre-optic light, whitish matter of very fine consistency can be seen to form bluish-white swirling veils running randomly through the rays of the star. Similar whitish veils have been observed in Czochralski pulled rubies, so the Kyocera star stones may have been manufactured by crystallization from a high-temperature melt by pulling, rather than by a flux-melt or hydrothermal process. Since there are no curved striae the Verneuil process is ruled out. One stone examined by the GIA showed a crude hexagonal pattern through the dome: this may have been the remains of a seed. Gas bubbles and swirls (seen as dark-edged wavy bands in shadowed transmitted light) are particularly characteristic of Kyocera stars.

As always there are no solid mineral inclusions, so in general the Kyocera stones should not cause too much trouble in identification. Company literature issued at the time of first release of the stones on to the market stated that two grades, A and B, would be made. The A stones would be near-perfect while the B stones would be very slightly flawed. Optical and physical properties are not distinguishable from natural ruby, but an internal examination should show enough features for the stone to be correctly diagnosed. The strong response to UV would be unusual though not absolutely impossible in a natural ruby, so readers are advised not to rely upon the admittedly pleasing UV test but to consult the microscope first.

So far we have looked at ruby and sapphire grown by the cheap and efficient Verneuil process and at the more expensive and slowly grown flux stones. Both types have characteristic inclusions which are by far the best means of identification.

Since ruby lasers are so important it is not surprising that growth of inclusion-free ruby rods suitable for laser use has also produced gem-quality material. Growth by pulling, in which a seed crystal is lowered to the melt surface and then slowly drawn upwards, taking the melt with it to form a rod, is not expensive and crystals with defects so slight that they affect laser performance can be used for gemstones instead.

Pink sapphire

In 1995 Union Carbide announced the growth of 'pink Ti-sapphire'. Other producers, including Kyocera and Russian enterprises, have also produced pulled corundum. The Union Carbide material is grown for gem use rather than as a spin-off from industrial work, and several other colours are on offer.

Pink corundum provides a nomenclature problem since if it inclines to red, sellers may call it ruby. Most gemmologists admit the existence of pink sapphire which is not merely a pale ruby. The name 'Ti-sapphire' was adopted by the GIA from a term commonly used in the laser context: they report on the pulled corundum in the fall 1995 issue of *Gems & Gemology.*

The pink sapphire to which titanium has been added has considerable laser importance so that the producers have two potential outlets for their product. It also means that research and development efforts will have financial backing.

An examination of faceted stones and rough material showed that colours were slightly orange–pink and purplish pink. The more saturated colours were said to have been annealed in a reducing (oxygen-free) atmosphere after growth. Compared with other synthetic pink sapphires and with natural pink stones, the titanium-doped and annealed, saturated colour Union Carbide product shows only a faint red fluorescence under LWUV and a weak to moderate blue colour with a slight or strong chalkiness under SWUV. Such a response is not seen with those natural or synthetic pink sapphires which are coloured by added chromium.

With the spectroscope the annealed stones show a weak absorption spectrum, and the as-grown specimens no absorption bands at all. Natural pink sapphires give a characteristic chromium absorption spectrum, and the same spectrum is shown by Verneuil-grown pink stones and by pink flux-grown and chromium-doped pulled crystals. Gemmologists would expect a pink sapphire to give a recognizable chromium spectrum and would be puzzled by the faint bands seen in the annealed Ti-sapphires. No natural sapphires coloured only by titanium have been reported.

Under magnification the Union Carbide sapphires were found to show minute high-relief bubbles and pinpoint inclusions. Pulled stones in general are likely to contain gas bubbles – these are sometimes elongated. When some of the annealed sapphires were immersed, some very elusive colour banding could be seen: di-iodomethane is a useful immersion liquid, and examination of the specimen by polarized light is also helpful. Such zoning was not seen in all the specimens, however. No curved colour banding could be seen, though this effect has been reported from some other pulled sapphires.

The Ti-sapphires could present a lot of problems to the gemmologist, and distinction between them and an admittedly rare but possible inclusion-free natural specimen might be beyond normal gemmological testing. Where the unusual fluorescence is encountered, further tests should be attempted or the stone sent to a laboratory. Unusual fluorescence in conjunction with the lack of an expected chromium absorption spectrum should lead to early recourse to the microscope to see if tiny bubbles can be identified.

While Verneuil, flux-melt and pulled crystals of corundum are all well known, the hydrothermal growth method has not been used to any great extent. The method is expensive compared to Verneuil and flux-growth techniques, and the use of a pressure vessel has its own problems. None the less there are some hydrothermally grown rubies around, grown on a seed whose coating usually contains profuse gas bubbles. The seed can be seen as a whitish area beneath the purplish-red coating: I have seen a crystal grown in this way from which platinum wires protruded – not usually seen with natural rubies! This specimen was an early experiment by Pierre Gilson whom we shall meet in the chapters on emerald and opal (Chapters 9 and 11).

Induced inclusions

While ruby crystals have been made by many methods, manufacturers have also found ways of making their product look more natural by introducing inclusions reminiscent of the ones found in natural stones. This is not a very widespread practice since the returns can only be small, but gemmologists should be aware that it goes on.

Natural corundum while containing a variety of mineral inclusions also shows flat planes of liquid droplets long known as 'fingerprints'. While some of the veils of flux seen when crystals are grown by the flux-melt method can resemble fingerprints quite closely, the veils are commonly twisted. Chatham and Knischka have reported to the GIA that they have induced fingerprint inclusions into rubies used for the study of crystal growth. Lechleitner, on the other hand, is reported to have introduced the fingerprints into synthetic rubies which are then placed on to the market.

In general the process appears to involve heating the specimen and then cooling it suddenly by placing in a liquid or melt. This thermal shock induces fractures within the crystals. They are then immersed in a flux melt containing dissolved aluminium oxide. The surface-reaching fractures take up the flux, which slowly cools within them, thus creating fingerprints which are made up of flux fragments. It is possible that similar features could be found in Verneuil crystals on first growth without further heating and quenching being necessary before placing in the flux. The process produces a layer of flux-grown material over the Verneuil material, which may be a complete boule or a small preformed stone. The flux material has been found to show parallel growth lines and some distinct crystal faces. When the final product has been faceted, the flux-grown coating may have been removed.

When some of the flux-grown material remains, the GIA have noticed some small doubly refractive crystals occurring only in the boundary between the coating and the core. From their RI they have been determined as corundum crystals with a different orientation from their host. The Verneuil seed has been found to contain several different types of fingerprint inclusions with a close resemblance to healed fractures in natural ruby. Gas bubbles seen in curved growth layers prove the Verneuil origin. Laboratories have found traces of molybdenum and/or lead, and these elements must have originated in the flux. They cannot be detected by gemmological tests.

In a paper published in the spring 1994 issue of *Gems & Gemology* the GIA report that flux-induced inclusions had been found in approximately 60 synthetic rubies that had been offered in the trade as natural. When viewed with reflected light and magnified, each stone showed a net-like pattern of whitish to colourless material reaching the surface. By transmitted light these inclusions form a thin continuous three-dimensional cellular structure resembling a honeycomb.

Gemmological properties of the inclusion-induced synthetic rubies are all within the normal corundum range, so the gemmologist has to rely upon the microscope for identification. Net- and honeycomb-like structures are probably the most distinctive signs to look out for.

The heat treatment of ruby and sapphire

While we shall not examine the techniques used in the heat treatment of ruby and sapphire in detail, we must always bear in mind that today any ruby or blue or other coloured sapphire may not be displaying the colour it was mined with. Questions of disclosure are also dealt with elsewhere, but the main question has always to be: is the colour stable? In the case of ruby and sapphire the answer will be in the affirmative, though colour applied by diffusion techniques may be polished away.

As a simple introduction to the kind of effect heat treatment may have, we can begin with looking at the finest rubies which were invariably associated with Burma, now Myanmar. For convenience we shall keep the name Burma ruby: Siam ruby is now generally known as Thai ruby, so possible geographical and political adjectives can be limited to one. The original Burma ruby, rich in chromium and poor in iron, has a unique red colour with a hint of purple – the name ‘pigeon’s blood’ has traditionally been used for these stones. Siam (Thai) rubies traditionally were a darker red, still very attractive but with a hint of brown. The Thai stones are now much less commonly seen in the gem markets than they were some years ago: of course they turn up in dealers’ old stock and in jewellery, but all agree that the familiar Thai colour is less common.

The reason for this is acknowledged to be the proved technique of heating the Thai stones to eliminate the brown component and produce a finer red. Today the gemmologist can fairly easily work out where the stone has really come from. ‘Burma’ colour is so important commercially that the catalogues of the major auction houses customarily state when a ruby is proved to be of Burmese origin. A fine Burma ruby of natural colour and good weight (10 ct would be a very good weight) commands the highest price of any gem species.

When examining any ruby of fine colour the gemmologist needs first to establish whether it is natural, synthetic (or some other natural stone or a glass); only then, when it has been proved to be natural will the question of provenance arise. If the ruby has come from Burma it will contain the natural mineral inclusions calcite or dolomite rhombs, hexagonal crystals of apatite and octahedra of spinel. Furthermore the Burma ruby will contain fine needles of rutile (the mineral whose crystals form the rays of stars in star ruby and sapphire when appropriately positioned). The needles intersect in a highly characteristic pattern where they meet at 60 and 120°: this familiar pattern is known as ‘silk’.

This pattern is not seen in the Thai stones since they contain no rutile: instead they very often show sets of long parallel lines arising from a crystallographic phenomenon known as twinning. The lines are not too difficult to see, and sometimes have tiny crystals strung out along them. This effect is absent in Burmese stones. The gemmologist who has studied photographs of inclusions must become familiar with these two important interior scenes which are not much altered by heat treatment. A ruby with ‘Burma’ colour which contains no rutile but does show the parallel lines is very likely indeed to be a heat-treated stone: once this is suspected (or even before), the gemmologist should look carefully at the facet edges and at the girdle. One of the effects of heat treatment is to cause pitting in these areas and a large stone may show them.

Geuda sapphires

Such a transformation is not confined to the ruby variety of corundum. A good deal of opaque to translucent whitish corundum can be found in Sri Lanka and was long disregarded as a source of gem-quality sapphire. At this point the story in brief is that Thai dealers or gemmologists found that some of this material, when heated under certain conditions, turned to a magnificent and stable blue.

The whitish (sometimes pale-blue) corundum is commonly if loosely known as 'geuda' and has to contain sufficient titanium for the blue colour to be brought out on heating. The name 'geuda' has a specific meaning in Sri Lanka, signifying translucent corundum of different colours, containing rutile as 'silk'; the name 'ottu' is used for clear corundum of different colours and 'dot' for opaque corundum. These names are explained by Harder in *Zeitschrift der Deutsche Gemmologischen Gesellschaft*, Vol. 39, p. 73 (1990), and elsewhere. In all cases the right amounts of iron and titanium have to be present for a fine blue colour to be developed. Though different authors vary in their views, temperatures of around 1600°C are said to be needed for the transformation to take place. Nassau in the second edition of *Gemstone Enhancement* (1994) suggests that stones are first trimmed or preformed to remove all but the smallest of inclusions and may then be immersed or painted with a borax-based solution. They are placed in an aluminium crucible and surrounded by charcoal or otherwise arranged so that a reducing (oxygen-free) environment is secured. The range of suggested additives proposed to reduce cracking, heal existing cracks and ensure a satisfactory colour change is large and sometimes contradictory. The period of heating is also variously reported, ranging from hours to days.

When the first heating appears to give promising results the stones may be heated again. If the results are unsatisfactory, modifications to the process may be made. Similar arrangements are made in other countries when sapphires are heated. Montana sapphires in particular (excluding the fine blue stones from the Yogo mine, whose colour as mined needs no improvement) occur as pale-blue, pale-green or pale-yellow crystals which, when heated, give a fine strong colour. Blue sapphires from Australia are characteristically so dark a blue that they may appear black: heating may lighten the colour. The geological history of Australian corundum and that of Sri Lankan stones is different, and variations of treatment are required.

The author was able to spend some time in the Montana sapphire deposits a few years ago and can confirm, through recovering a great number of differently coloured crystals, that heat treatment makes a great deal of difference to the appearance of the finished stones. The final colours can be quite aggressively strong, and are excellent for jewellery.

As we have already seen, some Verneuil rubies with curved growth striations may on heating have their effect made much less obvious by heating. Yet another process involves fracture inducement by thermal shock, stones (usually ruby) being heated to about 1000°C and then dropped into water, sometimes liquid nitrogen, according to Themelis in *The Heat Treatment of Ruby and Sapphire* (1992, published by the author).The resulting cracks can be filled with a dye or be covered with an overgrowth. The practice is intended to simulate natural fractures and inclusions.

Another process is the diffusion of colour into corundum, though the colour occupies only a thin surface layer. Both faceted and cabochon blue sapphires have been reported: very high temperatures are needed for the process to produce a coloured layer of tens of micrometres. Repolishing the stone for any reason will remove most if not all of the layer. Asterism may also be produced by diffusion, providing titanium is present. Blue is the easiest colour to diffuse into corundum but the *Australian Gemmologist*, Vol. 17, p. 326, (1990), reports on a suite of jewellery consisting of chromium-diffused orange synthetic sapphire: the red of ruby is less easy to diffuse since about 1 per cent of chromium is needed for distribution through the whole of the stone compared to the hundredth of a per cent of iron and titanium impurities needed for a good blue (Nassau, 1994).

The identification of heat- and diffusion-treated corundum has to start with the establishment of the identity of the specimen as corundum. For this the usual gemmological tests of RI, absorption spectrum and pleochroism, perhaps in extreme cases SG, can be used.

These tests will not normally show whether or not a ruby or sapphire is natural or synthetic, much less whether or not it is a treated natural or synthetic specimen. Once this has been established the lens or microscope has to be used. Heat-treated rubies and sapphires may show pitting of facets, damaged facet edges and girdles: these arise through the necessity for repolishing the stone after treatment to remove marks left on the surface by the high temperatures involved. Other signs of treatment are 'exploded' inclusions (hard to distinguish if you are not sure what they should look like anyway!): stress fractures like haloes or small discs surrounding solid inclusions are the tell-tale signs. A report by Koivula in *Gems & Gemology*, Vol. 22, p. 152 (1986), states that if intact inclusions containing carbon dioxide are found to be present the specimen cannot have been exposed to high temperatures.

Blue sapphires which have had their colour enhanced usually do not show the absorption band at 450 nm and may also show unusual pleochroic colours of violet–blue and greenish blue instead of the usual dark and light blue. Seen under UV, heat-treated blue sapphires often give a chalky green fluorescence and with the microscope will show no rutile (silk). The colour distribution may be diffuse or patchy. All these features should be looked for, and the diagnosis of treatment made on finding some or all of them.

When profuse rutile has been lessened by heating and when the brownish component has been removed from Thai ruby (as described above) there are no distinct signs of treatment: the natural inclusions in the stone will still be recognizable as characteristic of Burma or Thai ruby. When the rutile content has been diminished, the gemmologist will have to be content with establishing the natural or synthetic nature of the specimen.

Diffusion-induced stars need high temperatures, so the stones show the typical signs of heat-treatment. Diffusion colours may quite conveniently be seen if the specimen is immersed or even merely held against a white background: look for thin layers at the surface and for an uneven distribution. A diffusion-treated faceted stone may have lost the colour from some of its facets or from the girdle area. Where the stone shows surface pitting, colour may be seen concentrated at the bottom of the depressions, at a depth to which the polishing process could not reach. Star stones which have had their stars diffused in, show notably strong stars as well as colour-filled pits and edge damage. Nassau (1994) mentions that heat-treated corundum stones 'plink' against a hard surface while unheated ones 'plonk' (communication to Nassau from Robert Crowningshield of the GIA). This test should not be the basis of a final diagnosis!

Other colours of corundum

While ruby and blue sapphire are the most important varieties of corundum there is quite a strong demand for the bright-yellow stones, which can be very beautiful. We have seen the methods of synthesis and the ways in which ruby and blue sapphire are treated. The other colours (apart from the rare flux-grown orange stones) are usually grown only by the Verneuil flame-fusion method and do not have to be looked at again in the growth context.

Natural and synthetic yellow sapphires are in fact one of the banes of the gemmologist and gem-testing laboratory. Nassau (1994) lists seven types of yellow sapphire, drawing in this text from a paper by Nassau and Valente in *Gems & Gemology*, Vol. 23, p. 222,

(1987). Using the terms *yellow stable colour centre (YSCC)* and *yellow fading colour centre* (YFCC), the nature of which are outside of the coverage of this book (in practice, sapphires with the YSCC will not lose their colour on exposure to light while those containing the YFCC may do so over a period of time), Nassau shows that yellow sapphires do not always betray the stability of their colour by any other significant feature, nor do they always show whether their colour is natural or artificial.

Of the seven types of yellow sapphire described by Nassau, some natural stones have a stable colour which needs no artificial enhancement. This class includes the well-known Sri Lanka, Thai and Australian sapphires. Some Verneuil-grown synthetics also have a stable colour. Trouble arises with those specimens whose colour is derived from the irradiation of originally colourless material: some of these will slowly lose their colour on exposure to light. Specimens with surface-induced colour are also known. While there are occasionally reports that yellow heat-treated sapphires (rather than irradiated ones) have faded, this is probably a misunderstanding or misreporting.

There is no sure way of telling whether a yellow sapphire has the YSCC or the YFCC without actually subjecting it to a fade test. A few hours in sunlight are all that is needed. While the colour can be restored by further irradiation it will inevitably fade once more.

While ruby and blue sapphire can be altered in the various ways shown above, from the mid-1980s a further hazard awaited the gemmologist. We have seen how surface-reaching fractures in diamond can be filled with a glassy material: the aim is to disguise cracks and raise the clarity grade of the specimen. Oils have been used for the same purpose. It is likely that the stones are placed in a vacuum and the oil then added. In Thailand, bottles of 'ruby oil' are available! Glass infilling of fractures is not easy to distinguish – as always the gemmologist has to keep a suspicious mind when testing – but any prominent inclusion should be examined under the microscope. The presence of gas bubbles in an inclusion indicates filling, and when the specimen is immersed, added colour will be seen in parts that have been oiled. If plastic filling is used, its whitish colour will show up. A parcel of oiled corundum will give off an unmistakable oily smell, and the stone will feel tacky.

Imitations of ruby and sapphire

Ruby and sapphire can be simulated by a variety of materials ranging from simple glass imitations to quite convincing composites. Asterism (the star effect) can also be imitated.

Glass can always be detected by its internal swirliness, large randomly placed gas bubbles and fracturing on girdle and facet edges. Its RI will usually be in the range 1.50–1.70 compared to 1.76–1.77 for corundum: it will show no birefringence and no chromium absorption spectrum. Between crossed polars on the polariscope, glass will not show the normal pattern for anisotropic materials but an anomalous effect which gemmologists quickly learn to recognize.

Garnet-topped doublets may imitate blue sapphire or ruby: they are made almost entirely of glass but have a thin slice of almandine garnet (the dark-red mineral) fused to the table facet. This gives a red flash in strong light, and while this would probably pass unnoticed in ruby, it will certainly show in blue and the other colours of sapphire – though it is surprising how often the red flash is overlooked. In ruby, whose value ought to mean that careful checks are invariably applied, examination will show rutile needles in the garnet slice and confined to it: similar needles may of course occur in ruby but they will occur more widespread throughout the specimen. Should the specimen be immersed, the difference between garnet and ruby colour will at once be apparent. Gemmologists examining the absorption spectrum of a supposed ruby may encounter the strong bands of the almandine spectrum, which is quite unlike the chromium spectrum of ruby. The almandine absorption spectrum is also unlike that of blue sapphire, but here the red flash should give the deception away.

Many composites of ruby on ruby could be made, but in fact this practice is far less common than with emerald. Overgrowth on a seed is more likely to be encountered, but then the signs of flux growth should be apparent. The practice of foiling deceives many with the corundum gemstones, as it does with other coloured species: since the material used is usually glass the suspicious mind is the best test. Rubies and sapphires with induced stars show concentration of colour in pits and fractures and other signs of colour diffusion: Verneuil star stones have too perfect a star and can be identified by normal gemmological routines. Star corundum whose asterism arises from an engraved mirror placed at the back of the stone can sometimes be found in older jewellery: even the back of the setting can have star-like rays engraved on it. If all else fails the whole stone can be dyed, but any overall coating of this kind will not survive years of wear. Most specimens of these types are glass: however, one patented process has the rays of the star engraved on the back of the specimen itself, which will be a natural or synthetic ruby or sapphire.

Plastic coating of corundum (even with nail polish) is still found though 'aged' specimens treated in this way will show cracks in the layers. Acetone or the more simply acquired nail polish remover will remove the coating. Surface coating such as the 'Aqua Aura' blue-coated quartz has been reported on sapphire, but this gives a tarnish-like surface which should immediately indicate that something is wrong.

Simple gemmological tests serve only to distinguish natural from synthetic corundum and from other species. The microscope is the entirely essential tool for those gemmologists needing to identify all types of corundum: since very high prices are paid for fine specimens, practice in the use of the microscope should be the gemmologist's highest priority, and we are all caught out sometimes.

Cases from the current literature follow. Remember that while you are reading the book someone, somewhere, is devising a new process to confound you!

Reports of interesting and unusual examples from the literature

Items in this section have been chosen to illustrate points made in the chapter and to bring one-off items to your notice.

A Verneuil ruby which had probably been heated and quenched in a liquid, allowing a substance to enter the cracks and form 'fingerprint' inclusions is reported in the *Journal of Gemmology*, Vol. 23(2) (1992). The stone, containing gas bubbles from its flame-fusion growth, could have been very deceptive.

Ramaura ruby crystals have been identified either as single crystals or as penetration twins, the differences being described in the *Journal of Gemmology*, Vol. 24(2) (1994).

Pink sapphires grown by Chatham Inc., of San Francisco, California, are described in the *Journal of Gemmology*, Vol. 24(3) (1994). Specimens are a lighter shade of the firm's flux-grown ruby, and show a grid-like flux inclusion structure and also a strong orange–red fluorescence under both types of UV. Large flux inclusions near the surface gave a green fluorescence while others deeper in the stone glowed yellow.

Sometimes synthetic rubies have been fracture filled and any stone showing curved colour zoning, gas bubbles and other signs of flame-fusion manufacture may also show the characteristic 'foreign' flashes of orange or blue colour from filled fractures. *If the flashes are the first things to be noticed, a hasty decision on purchase might be made, since the ruby might be considered to be natural.*

In the November/December 1995 issue of *Colored Stone*, gem exports from Thailand are reported to have dropped 22.8 per cent, with ruby exports in particular down 22 per cent. Japanese traders were said to have complained about the number of treated rubies emanating from Thailand, and several parcels of rubies and sapphires, when intercepted by the Thai Gem and Jewelry Traders Association before export, were found to contain silica-treated stones.

An advertisement by the True Gem Company in *Colored Stone* for March/April 1995 states that their product of ruby and sapphire begins with the crushing of natural rough into pea meal size. Heavily included material and foreign substances are removed, and the residue heated to burn off impurities. Purification proceeds until gem quality is reached, when the material is melted and pulled to produce facetable crystals: the complete process is reported to take 6–8 months. The company claims that the presence of trace elements is enough to distinguish their product from those of other manufacturers of synthetic corundum. TrueRuby was being sold (1995) at US $230 per carat, and TruePinkSapphire and TrueBlueSapphire for US $230 per carat. Each stone weighing 0.5 ct or more has a registration number engraved on the girdle by laser.

Natural star corundum has been treated with a red dye, which concentrates in cracks. The aim is to make near-colourless or yellowish material resemble star ruby. Acetone removed some of the dye, and the stones gave no fluorescence.

Asterism induced by diffusion into the surface of two stones is reported in the fall 1985 issue of *Gems & Gemology*. One stone was a dark-blue cabochon of more than 40 ct, determined as natural corundum on the basis of its inclusions. *However, the stone gave a chalky greenish-white glow under SWUV and lacked the iron absorption line at 450 nm. The needles causing the star appeared, on immersion in di-iodomethane, to be confined to the surface and the same effect was seen in a light-red 6 ct cabochon of star ruby.* The ruby was found to be a synthetic flux-grown stone, and the needles were shown to be much coarser than those usually seen in synthetic star corundum. The thin diffused layer could easily be removed on repolishing.

A red stone simulating ruby and reported in the winter 1984 issue of *Gems & Gemology* turned out to be a composite with a colourless synthetic spinel crown cemented to a pavilion of Verneuil ruby. *Bubbles could easily be seen in the cement joining plane and curved striae in the ruby.* As the GIA remark, it is hard to see why such a specimen should have been manufactured.

A simulated ruby has been found whose colour arises from a red material lining the drill hole. This practice can deceive when found (as it usually is) in very small stones.

In the January 1994 issue of the *Journal of Gemmology* the danger of confusing a filled fracture in ruby with a natural inclusion is pointed out. In a cabochon ruby a large inclusion occupied a large area of the base of the stone and showed a whitish cast unlike the effect shown by some types of filling. The material turned out to be calcite, a frequent inclusion in natural ruby. Identification was carried out using a small amount of dilute hydrochloric acid.

A blue sapphire strongly resembling a Sri Lankan native-cut stone seen by the GIA and described in the summer 1987 issue of *Gems & Gemology* had an unusual depth-to-width ratio of approximately 115 per cent, making it more deep than wide. It was apparently inclusion-free, and the colour was good. However, under magnification curved colour banding could be seen near the culet, and stress cracks were apparent on some facets. Under SWUV there was a strong chalky blue fluorescence suggesting artificial origin or heat-treatment at a high temperature. The Plato test identified it as a flame-fusion synthetic. It is wise to be careful with any stone of unusual proportions.

In the spring 1987 issue of *Gems & Gemology* a zone of brown colour in a heat-treated blue sapphire is reported. This is not a customary feature of heat-treated stones. Lack of distinctive bands in the absorption spectrum suggested that the stone came from Sri Lanka rather than Australia, whose iron-rich sapphires absorb strongly. It is possible that the stone was originally a geuda (whitish translucent Sri Lanka sapphire) which sometimes shows brown stains on heating.

In the spring 1986 issue of *Gems & Gemology*, Nassau warns against a fade test for yellow sapphire in which specimens are heated to 200°C for one hour. It is possible that stones may lose their colour.

Doublets of natural black star sapphires are reported in the fall 1985 issue of *Gems & Gemology*. One specimen showed a separation plane of colourless and nearly transparent cement joining the two portions. This layer melted easily with the thermal reaction tester.

Sapphires from Montana and Sri Lanka are not the only ones to have been heated. In the *Journal of Gemmology*, Vol. 20(7) (1987), characteristic signs of treatment are reported in blue sapphires from the Umba River area of Tanzania. *Coloured haloes surrounding mineral inclusions are normally accepted as proof of treatment.*

Brown synthetic star sapphire is reported in the fall 1988 issue of *Gems & Gemology*. This is a rare colour: the stone showed no absorption in the visible region and was inert to both types of UV radiation. With some semi-transparent areas near to the surface, the stone was otherwise opaque, and *widely separated curved growth bands could be seen.*

One consequence of heat-treating rubies is that *shallow depressions may be formed on the surface, some of the depressions showing signs of partial melting. Most cavities are fractions of a millimetre in diameter, ranging up to 1 mm. Since polishing them away would lessen the weight and hence the value of the stone, some rubies may still show these distinguishing features when placed on the market. Sometimes the cavities have been filled with a glassy material so that the surface looks less pitted.*

A fine orange–yellow faceted sapphire of 7.60 ct was found to have been heat treated, *the signs being three zones of chalky blue fluorescence close to the girdle, seen under SWUV. Immersion in di-iodomethane showed straight angular yellow coloured zoning alternating with near-colourless areas (characteristic of a natural corundum) and three straight blue zones close to the girdle, coinciding with the fluorescent areas.* According to the description in the fall 1987 issue of *Gems & Gemology* these features could have resulted from heat treatment.

The ruby crystals produced by Professor P. O. Knischka of Steyr, Austria, have yielded faceted stones up to 67 ct in weight, according to their grower. In 1990, ruby was being manufactured as faceted stones up to 11 ct and as preforms which may be larger than 25 ct. Ruby is also offered as macroclusters, plates and microclusters. Colours range from 'Thai' through light, medium and dark Burmese pink. In the fall 1990 issue of *Gems & Gemology* the GIA describe a large half crystal of the Knischka ruby. The crystal had been sawn down its length, and was 39.99 mm long and 17.99 mm across its largest diameter. The weight was 40.65 ct. The colour was a dark purplish red with accordion-like deep growth steps perpendicular to and along the length. Glassy two-phase inclusions and platinum platelets could be seen.

In 1989, an *ICA Laboratory Alert* was issued on two examples of synthetic corundum placed on sale as natural water-worn rough. The larger crystal weighed 56.04 ct and was purple in daylight and purplish-red in incandescent light. The smaller crystal weighed 21.80 ct and was medium blue, resembling a broken and worn sapphire pebble. Both crystals showed the characteristic signs of flame-fusion growth, with concentric colour banding. The larger crystal showed 'steps' on the surface; the smaller crystal had step-like elevations and depressions on its surface. All these features were consistent with the use of a grinding wheel. The two crystals had been purchased in Sri Lanka.

While most gemmologists would consider Verneuil flame-fusion corundum to be easily identifiable, some specimens can cause confusion until a diagnostic feature is found. *Needle-like inclusions are sometimes seen, and even straight twinning. Such features have been reported from blue and orange sapphire and from ruby.*

Triangular inclusions in Verneuil flame-fusion blue sapphires are noted in the spring 1991 issue of *Gems & Gemology.* While most inclusions in this type of product are round, the ones described were in fact triangular cavities containing gas bubbles.

A blue star sapphire in which both colour and asterism were produced by surface diffusion is described in the spring 1991 issue of *Gems & Gemology. Immersion in di-iodomethane showed the diffusion-produced colour, and the star effect was weak.*

A flux-grown 3.5 ct orange–red synthetic ruby described in the spring 1991 issue of *Gems & Gemology* showed unusual near-colourless straight parallel bands sandwiched between areas of orange–red colour. This effect could deceive the unwary into mistaking the specimen for natural ruby.

Some synthetic ruby crystals have been cut to resemble the hexagonal crystals of red beryl. Characteristic inclusions of a flame-fusion ruby (curved striae, gas bubbles) eliminate the possibility of red beryl, although prism faces are artificially abraded.

Occasionally quite simple techniques are used to improve the visual quality of stones. Cutters in Sri Lanka have been known to rough-grind the pavilion facets of good-quality blue sapphires: the ground faces increase light scattering and give a velvety appearance, similar to that shown by Kashmir blue sapphires. Some synthetic ruby and spinel cabochons are given rough bases to increase their 'natural' appearance.

Without investigating the meaning of the name, 'Geneva rubies' are still turning up in jewellery. The name is commonly used by older jewellers and dealers for early Verneuil synthetic rubies, which are characterized by very tightly curved striae, prominent gas bubbles, colourless areas and strain cracks, with some black inclusions said to be undissolved alumina. The stones date back to 1866 when the French Syndicate of Diamonds and Precious Stones ruled that they had to be sold as man-made.

A synthetic ruby, perhaps an early product, showed a blue outline encircling the seed, which was natural corundum. In the summer 1991 issue of *Gems & Gemology* the GIA speculate on whether the presence of iron and titanium in the natural seed produced the blue colour. It is possible that heat from the growth process assisted iron and titanium to migrate to the crystal surface, where they would come into contact with the flux solution.

Materials with a colour close to that of the pink–orange 'padparadschah' sapphire include erbium-doped yttrium aluminium garnet (YAG) and erbium-doped lithium fluoride. *Both these materials will show a rare earth absorption spectrum.* The same firm (Synoptic) that manufactures these materials grows YAG doped with a combination of chromium, thulium and holmium to give an emerald green, which colour is also shown by chromium-doped lithium calcium fluoride, and chromium and neodymium-doped gallium scandium gadolinium garnet. *The first and third of these materials will show a rare earth fine-line spectrum.*

In the winter 1991 issue of *Gems & Gemology,* two synthetic sapphires with blue and orange colours respectively, were found to show twinning. This had not previously been reported for synthetic sapphires, and it had always been held that such an effect meant that the stones were natural. *Gas bubbles and curved growth lines established the identity of the sapphires beyond doubt as flame-fusion synthetics. The twinning could be seen as straight parallel lines. To add to the confusion, needle-like crystals of boehmite could be seen between the twinning lines.*

A medium yellowish-orange coloured sapphire weighing 0.98 ct was examined by the GIA and described in the winter 1991 issue of *Gems & Gemology.* Standard gemmological tests established the specimen as sapphire. No absorption features could be seen with the hand spectroscope, and there were no luminescent effects which would have indicated heat treatment. Though the colour of the stone was reasonably strong, no pleochroism could be seen, and this certainly would have been expected from a normal corundum. In reflected light a purple iridescence could be seen on the pavilion facets, and in diffused dark-field lighting colour was seen to be concentrated along the edge of the pavilion surface. Coating was the only explanation, and was confirmed after soaking the stone in concentrated hydrochloric acid for five hours at room temperature. This decreased the depth of colour though allowing some colour to remain on the pavilion. After two hours further soaking the colour had entirely disappeared. The orange colour was believed to arise from absorption in the green and blue portions of the spectrum.

Pink sapphire produced by the Czochralski pulling method owes its colour to doping with titanium, as reported in the spring 1992 issue of *Gems & Gemology. A strong orange–red colour could be seen through the Chelsea colour filter and a very strong bluish-violet fluorescence was observed under SWUV.*

Green and blue sapphires have also been produced by the same crystal growth method. Rods examined by the GIA have shown concentric colour zoning from the core to the rim, similar to that shown by Verneuil-grown sapphire. Stones are reported to be inclusion-free.

The very fine-quality rubies grown by Professor Paul O. Knischka of Steyr, Austria, were reported in 1992 to be on sale from the Argos Group of Los Angeles, California.

Synthetic star stones usually show so complete and central a star that suspicions are aroused at once. The GIA report in the winter 1991 issue of *Gems & Gemology* on a synthetic blue star sapphire in which the rays extended only a short distance into the stone from the periphery, leaving a blank central space. This central area showed greater transparency than the remainder of the stone. Such an effect is unlikely to be seen in a natural star corundum.

In the fall 1992 issue of *Gems & Gemology* a sapphire dyed to imitate ruby is reported. The stone, forming one bead from a necklace, *showed under magnification that the red colour was concentrated in cracks which penetrated the entire stone. No chromium absorption lines could be seen with the spectroscope but a fairly strong iron line at 450 nm was present, suggesting that the original material was sapphire.* This was confirmed when a single bead was sectioned, one half showing a broad band centred at 560 nm, and iron absorption at 450 and 380 nm, using the UV spectrophotometer.

It is well known to gemmology students at least that the curved striae in yellow Verneuil-grown synthetic sapphire are very hard to see under normal conditions. *Gems & Gemology* for summer 1992 *recommends the use of a filter of a complementary colour, thus blue for yellow or orange stones.*

The so-called 'Geneva rubies' are now known to be early Verneuil flame-fusion products. The summer 1992 issue of *Gems & Gemology* reports on a period ring containing nine such stones, the largest weighing more than 1 ct. *Their origin was determined by the presence of tightly curved growth lines and black inclusions. It is rare to find 'Geneva rubies' of such a size without prominent strain cracks.*

A transparent light-yellow sapphire weighing 8.27 ct was shown to be a natural stone by gemmological testing. *Heat treatment was proved by the presence of small discoid fractures around included crystals. While a moderate orange fluorescence under LWUV is characteristic for heat-treated and untreated yellow sapphires, this stone showed an atypical overall yellow fluorescence under SWUV. Magnification under a short-wave UV lamp showed that most of the stone glowed orange but some areas fluoresced a chalky bluish white. These areas were light-blue colour zones which may have developed during heating. The chalky blue and underlying orange colours combine to give the overall yellow colour observed.*

Diffusion-treated stones often show wear on crown facets, but some stones are so damaged that repeated heating may not be the only cause. A blue sapphire of about 4 ct set in a ring with side diamonds, reported in the winter 1992 issue of *Gems & Gemology*, showed microchipping at the edge of the table, similar to that shown by heat-treated zircon after years of wear. It is possible that the durability of the surface layer of a diffusion-treated stone is less than that of a stone which has been heat treated.

In the fall 1992 issue of *Gems & Gemology*, diffusion-treated sapphire cabochons in different colours are reported to have been sold as natural stones in sizes down to 3 mm. Many have been sold in the sapphire-mining regions of Thailand. The cabochons can be identified by the characteristic blue outlining of the girdle when the stone is immersed in di-iodomethane. Occasionally the detection is not apparent until the stone is repolished: then the turning of the stone against the lap produces an effect similar to a snail shell, the spiral pattern resulting from partial removal of the colour.

An imitation ruby crystal made from melt-grown synthetic ruby is featured in the fall 1993 issue of *Gems & Gemology*. The stone weighed 10.85 ct and was in the shape of a slightly distorted hexagonal pyramid with a polished base. The 'prism faces' had a water-worn appearance and irregularly spaced 'striations' sawn across them. *Tests showed that the specimen was in fact ruby, but the interior showed a network of small cracks which suggested that the piece had been heated and then quenched in a cold liquid. Some possible gas bubbles could also be seen.* Energy-dispersive X-ray fluorescence (EDXRF) analysis proved the stone to be a melt-grown synthetic ruby.

Hydrothermally grown stones, apart from the varieties of quartz, are not so common in the world of synthetics as those grown by flame-fusion or with a flux. From time to time someone makes a hydrothermal ruby – in the 1970s Pierre Gilson made a few – but on the whole they are rare. In the winter 1992 issue of *Gems & Gemology* a hydrothermally grown ruby crystal of 1930 ct is illustrated. Such crystals, with many other species of great ornamental attraction, were grown experimentally in the 1970s when all kinds of materials were being tested for their research or industrial potential. The report says that the ruby was grown in a concentrated potassium carbonate solution in a silver-lined vessel at a pressure of 20 000 psi and temperature of 540°C. Veils of fluid inclusions are prominent, and form a roughly honeycomb-like structure.

In the winter 1992 issue of *Gems & Gemology* the GIA report on a star sapphire weighing 7.08 ct which had to be classed as sapphire rather than ruby because the colour was a dark reddish purple. The high transparency shown by the stone was attributed to a low rutile content. Though a synthetic product, the stone resembled natural material more closely than most.

Enhancement of ruby is cited in the fall 1993 issue of *Gems & Gemology*. In the report, a 9.0 ct East African ruby cabochon was found to show octahedra in the glassy surface layer of a shallow pit on the base of the cabochon. Using X-ray diffraction techniques the 'glass' surrounding the octahedra was found to have an approximately zoisite composition while the octahedra had a spinel composition. An analysis taken round the rim of the pit showed a glass enriched with alumina and calcium oxide. The three components around the pit were thus identified as artificial glass, zoisite and spinel, with the last two minerals adhering as devitrification minerals of the glass coating. On the basis of the examination of other treated rubies, Dr Henry A. Hänni of the SSEF Swiss Gemmological Institute, who carried out the report discussed above, has wondered whether every glassy filling in corundum is produced with the intention of filling surface pits and cracks, or if boron or fluorine compounds are used to protect the stones during routine heat treatment. As the coatings melt they may act as a flux dissolving alumina and other compounds from the host corundum and then recrystallizing them in surface-reaching fractures.

Star doublets are mentioned in gemmology textbooks but are not very common. The GIA report, in the fall 1993 issue of *Gems & Gemology*, on a transparent red cabochon, bezel-set in a man's ring, and showing a six-rayed star. *The RI and other features showed that the crown was a Verneuil synthetic ruby and under magnification round and oval gas bubbles could be seen in the cement layer. The base of the cabochon, which was a reddish-purple colour, showed strong hexagonal growth zoning, partially healed fractures and other signs suggesting it to be natural corundum of the low-quality type sometimes known as 'mud ruby'*. The report goes on to say that a positive identification of the base portion was not possible, however, since the setting prevented it. The rutile crystals ('silk') in the base did, however, provide the star effect in the crown.

Despite the claims in gemmological textbooks, the curved banding (striae) seen in much Verneuil corundum is in fact not easy to see. Immersion is often the answer, and the all-purpose liquid di-iodomethane is one of the most easily available and generally useful liquids for the gemmologist. Verneuil rubies thus immersed and viewed with bright-field lighting ought to give up their secret. The GIA recommends the insertion of a green plastic filter between the objective lens and specimen to give colour contrast (in the same way, a blue filter can help in the resolution of bands in yellow sapphires). The combination of bright-field illumination and immersion has been used in the testing of Czochralski-pulled alexandrites.

Since the ruby deposits of Vietnam began to produce high-quality material it was clear that cheap imitations would quickly turn up on local markets. In the fall 1993 issue of *Gems & Gemology*, the GIA staff report the purchase at a Vietnam mine site of 'badly water-worn stones'. Four of the five stones were Verneuil ruby while the fifth turned out to be glass. *The latter was a convincing ruby imitation with a medium-dark purplish-red colour, giving an RI of 1.649 and an SG of 3.84. Strain birefringence was seen through crossed polars, the specimen showed red through the Chelsea colour filter and a weak chalky blue fluorescence was seen under SWUV while remaining inert to LWUV. Under the microscope the piece was found to contain two wedge-shaped layers of spherical gas bubbles. The spectroscope produced an absorption spectrum which included many distinct lines over the whole visible area, with broader, distinct lines between 590 and 570 nm.* EDXRF analysis proved that neodymium and lead were present in the glass, the neodymium clearly being responsible for the rare earth absorption spectrum.

Colour change in gemstones is by no means as uncommon as the gemmology textbooks appear to suggest. In the spring 1994 issue of *Gems & Gemology* the GIA report on a reddish-purple round brilliant set in a ring. Normal testing showed the stone to be corundum, and curved striae proved it to be a Verneuil synthetic. The absorption spectrum was characteristic of those colour change synthetic sapphires owing their colour to vanadium. However, on examination under both fluorescent and incandescent lighting, a red component could be seen. This was found to be caused by a red foil or paint-like coating on the pavilion facets.

During the 24th International Gemmological Conference held in 1993, some current-awareness topics were introduced. The Douros synthetic ruby was found to show growth structures very similar to those already encountered in the Ramaura flux-grown stones. Laser tomography was put forward as a possible means of distinguishing between heat-treated and naturally coloured rubies. With a laser beam as a light source, laser tomography can record on film the distribution of scattering centres (submicron-sized and larger inclusions) as well as growth banding. With this method the smallest scattering centres can be seen at low magnification. In an example shown, a ruby of 0.68 ct exhibited the curved bands characteristic of the flame-fusion product and a pattern produced by treatment-induced flux fingerprints resulting from overgrowth.

A diffusion-treated product exhibited at the 1994 Tucson Gem & Mineral Show consisted of small pale-blue synthetic sapphires (in the size range 0.60–0.75 ct). They had been diffusion treated with a cobalt compound, this giving a colour similar to that of some cobalt-coloured synthetic spinels. *The absorption spectrum showed three bands centred at about 620, 580 and 545 nm. There was no response to LWUV, but a weak chalky bluish-green fluorescence could be seen under SWUV. Stones showed a saturated dark red through the Chelsea colour filter: all these features were consistent with cobalt-doped sapphires. In addition, specimens under magnification were found to contain gas bubbles and/or curved colour banding, proving that the starting material was a Verneuil-grown synthetic.* Cobalt-diffused synthetic sapphires in which the starting material was a light pink were also seen at Tucson. They were produced as a simulant of tanzanite since they gave a cobalt-blue colour in fluorescent light and a violet to purple colour under incandescent light.

Sometimes confusion between coated and diffusion-treated sapphires can be found in the gemstone trade. According to a report in the spring 1994 issue of *Gems & Gemology*, a dealer in Bangkok was asked to examine a parcel of four small sapphires whose colour had been enhanced by a new 'electric-blue' diffusion process. Since one of the stones had a large broken surface revealing an essentially colourless stone with colour confined to a shallow surface area it was natural to assume that the stone was a diffusion-treated sapphire.

However, on immersion in di-iodomethane and study with a microscope, the expected dark colour outlining facet junctions was not observed. A dimpling of the surfaces was seen and a very slight repolishing of the facets rapidly removed the coloured layer. This showed that the stones were coated to give the colour rather than having it diffused into them.

Rubies are always the most likely gemstones to be synthesized as more than one process is available and because there will always be a ready sale for the product. In the spring 1994 issue of *Gems & Gemology* a Greek-made ruby is described. The Douros ruby was available in both cut and rough form.

The firm of Chatham, long celebrated for its emerald and more recently for its orange and blue sapphires, produced a pink sapphire during 1993, the stones being shown at the 1994 Tucson Gem & Mineral Show. The chromium content of 0.06–0.2 wt% compared to that of the ruby (0.5–2.0 wt%) gives the pink colour. *Properties are normal for pink corundum but the slightly orange–red fluorescence seen under SWUV is of about the same intensity as the red fluorescence under LWUV. This is not seen with natural ruby where the LWUV response is almost always the stronger.* The single quality of stone placed on the market is described as 'clean'. *The crystal examined by the GIA contained elongated flux inclusions parallel to striations seen on the crystal: the inclusions fluoresced green under SWUV only, but they may show yellowish when this colour is combined with the red body colour of the ruby.*

An update of diffusion-treated sapphires is given in the spring 1994 issue of *Gems & Gemology*. At the Tucson Gem & Mineral Show in that year there seemed to be fewer examples than in the three previous years. One anonymous exhibitor said that he was experimenting with red diffusion-treated stones. Another report said that diffusion-treated material turned up regularly in parcels of sapphires sold in Bangkok: the treated stones were 'salted' into the parcels. When some of the sapphires were repolished *the characteristic bleeding of the diffused colour into surface pits and fractures could be easily seen.*

Hydrothermal ruby is much less common than its flux-grown counterpart, but it may be seen more frequently in future. A news item in the fall 1994 issue of *Gems & Gemology* describes and illustrates a set of hydrothermal autoclaves in use at the Tairus facility in Novosbirsk, Russia. Emerald is also made at this site, which opened on 3 August 1994. Ruby and YAG in a variety of colours is also grown by a modification of the floating-zone technique, called horizontal growth.

While most synthetic sapphires made by the flame-fusion process are evenly coloured, the occasional stone may surprise the gemmologist. In *Gems & Gemology* for fall 1994, a 7.03 ct sapphire is reported to have shown normal properties for corundum and a uniform yellowish-orange colour face up. *When examined from the side, though, the stone appeared pink except in a small orange area along the keel line at the bottom of the pavilion. When the stone was examined with a blue colour-contrast filter with diffuse transmitted light, colour in the keel zone appeared as very weak, diffused and curved orange bands. The banding and clouds of gas bubbles proved the specimen to be a flame-fusion synthetic.*

Bright-yellow synthetic sapphires grown by the pulling method are reported in the summer 1994 issue of *Gems & Gemology*. Apart from ruby and blue sapphire, growth (other than by the flame-fusion method) is not usually undertaken for economic reasons. However, stones on view at the 1994 Tucson Gem & Mineral Show were a highly saturated slightly greenish-yellow colour. They resembled some varieties of golden beryl rather than yellow corundum, and were found to contain nickel but no other chromophore in significant amounts. *With the spectroscope, a faint chromium emission line at 690 nm could be seen, and the three stones tested fluoresced a faint orange under SWUV, one specimen showing a similar effect under LWUV. Microscopic inclusions, perhaps gas bubbles, and curved growth features were seen in all three stones when immersion, bright-field illumination and a blue contrast filter were used.*

Green star sapphire is very uncommon, and when the GIA examined such a stone in 1994 (reported in *Gems & Gemology* for spring 1995) the constants proved to be normal for corundum with the exception of an absorption line at 670 nm, which was attributed to cobalt. The colour from the photograph is an attractive light green, quite unlike most green sapphire, which is iron-rich. The star was caused by a thin, cloudy and mottled area just below the dome of the stone. This effect has been seen in other synthetic star sapphires. *Probably the unusual colour and the cobalt absorption line will be enough to alert the gemmologist*, but examination by EDXRF showed that cobalt did appear to be the colouring agent. Some green non-star sapphires have been found to contain trivalent cobalt and trivalent vanadium.

Corundum with diffusion-induced asterism began to appear in the gemstone trade during 1994, and a report in *Gems & Gemology* for spring 1995 describes a 23.26 ct purple star sapphire whose star appeared cloudy. *When immersed in di-iodomethane the stone showed a thin red layer apparently confined to the surface, and a number of red spots could be seen, accompanied by a red cloud, on the dome of the cabochon.* It is believed that the original intention was to make a diffusion-treated ruby and that impurities in the original cabochon may have caused a contamination of the colour, leading to a decision to create a star stone by titanium diffusion into the surface.

In the winter 1994 issue of *Gems & Gemology* the GIA report on a transparent pink 15.58 ct modified brilliant with properties established as those of corundum. Under magnification only pinpoint inclusions could be seen: *use of the Plato method finally proved it to be synthetic. However, under SWUV the stone displayed a strong but uneven bluish-white fluorescence (it gave the expected red glow under LWUV), this effect proving to arise from clearly defined colour bands. It might be well worth while to use SWUV on 'borderline' corundum varieties: Hughes (Corundum, 1990, Butterworth-Heinemann) makes the point that low magnification and SWUV can be used to locate growth details in synthetic colourless sapphires, adding that a short-wave blocking filter is essential.*

It is often useful to try a white diffusion filter to help the resolution of colour banding in blue synthetic sapphires, and a blue diffusion filter to do the same in yellow synthetic sapphires, in which the bands are particularly hard to see. A similar test is also useful for the detection of growth sectors in synthetic diamonds.

A colour change sapphire showing twinning lamellae is reported in the summer 1995 issue of *Gems & Gemology*. The 1.43 ct transparent emerald-cut stone appeared bluish green in daylight-equivalent fluorescent light and pinkish-purple in incandescent light. While no gas bubbles or negative crystals could be seen under magnification, the stone showed a series of laminated twin planes. This effect is normally associated with natural corundum, *but since curved striae were observed through the pavilion facets of this specimen it was clearly a Verneuil synthetic. Gemmologists should study all the internal features of a specimen before making a final identification!*

Although gemmological texts warn of synthetic gemstones with induced natural-appearing inclusions, they have not so far appeared in great numbers on the market. In the summer 1995 issue of *Gems & Gemology* the GIA report on a 21.28 ct star ruby of a 'Burmese' colour with a moderate star and apparently native cut with a very thick girdle. While the piece was easily determined as ruby, fingerprint inclusions could be seen with the microscope. Under higher magnification, however, curved striae could also be seen. Both fingerprints and star were surface diffused.

The Kashan synthetic ruby has become well known in the gem markets since its first appearance in the early 1970s. Gemmologists have become aware of and used to such signs of flux growth as 'breadcrumbs' or 'paint splash' inclusions. The firm has now embarked upon a programme which will grow rubies and sapphires (the latter not previously produced) manufactured by a 'solution-growth' process instead of what the company calls 'forced-growth'. All Kashan products will still be flux grown. In *Gems & Gemology* for spring 1995 the GIA report on an 0.84 ct round, mixed-cut sapphire from the higher-quality end of the new range. *The stone was a transparent purplish-pink colour with properties similar to those of the earlier material.* While chromium was present as expected, the stone contained a lower proportion of titanium than that found in Kashan rubies.

In recent gem shows, blue sapphire manufactured by the Czochralski pulling method of growth has been on sale. *Gems & Gemology* for fall 1995 reports that some sapphire crystals have been unevenly coloured with the first-grown portions lighter in colour than the remainder of the rod. This is an opposite effect to that seen in split boules of Verneuil-grown sapphire where the colour is lighter in the centre with a darker outer portion. *Faceted samples with normal corundum properties showed a faint pink through the Chelsea colour filter and a weak red fluorescence under strong light from a fibre-optic source. Such a reaction is not unknown in Sri Lankan blue-to-violet sapphires and in some violet–blue Verneuil-grown stones. Stones also showed minute gas bubbles, but slightly curved blue banding, proof of artificial origin, was observed with difficulty, specimens needing to be immersed in di-iodomethane for satisfactory observation.* Elongated bubbles, sometimes seen in pulled crystals, were not reported in this case.

A diffusion-treated corundum specimen was reported by the GIA in the fall 1995 issue of *Gems & Gemology.* Three transparent purplish-red oval mixed-cut stones similar to rubies mined in Thailand showed the optical properties of corundum. Under magnification the smallest stone showed an opaque, light greyish-blue silky zone with a hexagonal outline appearing to be the core of the stone, this proving it to be natural. A crystal inclusion had altered to resemble a 'cotton ball', and in the other two stones the facets showed uneven surfaces with small cracks, resembling a surface coating. With diffused light passed through the crown, the middle-sized stone showed all the pavilion facet junctions more strongly coloured. When the stone was immersed in di-iodomethane, concentration of colour along the facet junctions was easily seen. This effect proved the stone to be a diffusion-treated corundum.

Rubies and pink corundum marketed as 'Recrystallized ruby and pink sapphire' by the TrueGem Company of Las Vegas is examined by the GIA in the Summer 1995 issue of *Gems & Gemology.* The three stones examined had identification numbers lasered on their girdles and showed the normal properties of corundum. *Growth banding or curved striae could be seen in the samples.* X-ray fluorescence indicated the presence of aluminium, chromium, titanium, vanadium and iron. Gallium was also detected. The concentration of these elements was consistent with a pulled or Verneuil-grown product and not consistent with natural corundum.

Specimens of gadolinium gallium garnet (GGG) have been offered as water-worn crystals of natural ruby, *Gems & Gemology* reports in its winter 1995 issue. The crystal reported had been ground to give a rough surface and also showed parallel striations (grooving). By transmitted light the colour was a deep purple.

Another ruby imitation was cleverly made from inserting a blue waxy material into a cavity in a synthetic ruby (presumably a cheap Verneuil-grown stone). Since a blue central zone is characteristic of rubies from the Mong Hsu mines of Myanmar, it is possible that careless buyers may be taken in.

A sapphire examined by the GIA in 1995 and reported in *Gems & Gemology* for winter 1995 was near-colourless and weighed 4.57 ct. The properties were as expected for both natural and synthetic sapphire, and it was only when the stone was magnified while under SWUV that the characteristic curved growth lines of a flame-fusion synthetic sapphire could be seen. The interesting phenomenon was that after irradiation (for about three minutes) the stone had turned a medium brownish yellow. It took about six hours in a solar simulator for most of the colour to disappear: however, the flame of an alcohol lamp returned the specimen to its original condition in only a few minutes. Jewellers and dealers have many reasons to suspect any sapphire, even of the faintest yellow, as we have seen elsewhere in the book, but a fade test is better not carried out on a customer's stone! The absence of natural inclusions is probably the best test, using the 10× lens.

Chapter 9

Emerald and its relatives – the beryl group of gemstones

The beryl group of gemstones includes the celebrated emerald and aquamarine as well as golden and yellow beryl and a rare red variety. Unlike diamond (an element) and corundum (an oxide), beryl is a silicate: the significance of this is that man-made growth is more complicated though not impossible. Commercial considerations dictate, however, that only emerald is synthesized, though the other varieties are often imitated.

Emerald is not as rare as ruby, though indifferent specimens of both are not hard to find on the market. Geologically, emerald is far more commonly found than ruby: despite still being a rare and expensive gemstone, there is never an emerald shortage as there often is with certain sizes and qualities of ruby. The chemical composition of emerald as a silicate means that the simple Verneuil method of flame-fusion growth cannot be undertaken for technical reasons, so that all synthetic emeralds have to be grown by the flux method or hydrothermally. As these methods are expensive to run there are no really cheap synthetic emeralds as there are cheap synthetic rubies and sapphires. In recent years a synthetic aquamarine has been made in Russia, but at the time of writing in mid-1996 there is no commercial exploitation of this material.

As with corundum, synthetic emerald has been around much longer than most people think. Early experiments on growth date back at least to the 1880s, and the early products were to some extent fairly easy to identify. By the middle of the twentieth century, synthetic emeralds were beginning to cause problems to jewellers and gem-testing laboratories, and one of the early gem-testing instruments, the Chelsea filter, was devised at this time with the intention of separating natural from synthetic products. This simple filter, transmitting either red or green light, enabled gemmologists to distinguish between natural and synthetic emeralds (for the most part) on the basis of their chromium content. Early synthetic emeralds transmitted a brighter red than most natural specimens.

For a green beryl to earn the name of emerald it must contain chromium, whether or not it looks like the common conception of emerald. If no chromium can be found in a specimen it must be called green beryl even if, as sometimes happens, the appearance is exactly the same as emerald. This ruling is enforced by international jewellery and gemstone organizations.

As synthetic emerald has the same composition as its natural counterpart the normal gemmological tests will not usually distinguish natural specimens from synthetic ones. Some synthetic emeralds have a specific gravity (SG) as low as 2.65 (the same as for quartz) but such stones are quite rare, the usual SG range for synthetic emerald being closer to 2.66–2.69 compared to 2.66–2.80 for natural stones. For synthetic emerald a characteristic refractive index (RI) range would be 1.56–1.58, overlapping effectively with most natural emerald, though a few natural stones reach 1.60. Some early synthetic emeralds may show a birefringence as low as 0.003 (compared to 0.006) but neither the SG, RI or DR (double refraction or birefringence) can give more than a first inkling that a stone may not be natural. We may state at this point that most synthetic emeralds will transmit ultra-violet radiation down to about 230 nm whereas most natural stones are opaque to wavelengths of 300 nm or lower. This may not apply to some recent productions.

In theory most synthetic emeralds, from whose growth environment iron can be excluded, should give a bright-red fluorescence under long-wave UV: in practice stones showing this effect are not often seen and the Chelsea filter or crossed filter test is cheaper and gives better results. The crossed filter test simply involves the irradiation of a specimen with blue light (such as that provided by a photographic or other filter, or by a flask of copper sulphate): the effect is then viewed through a red filter, through which chromium-rich emeralds glow a bright red. Natural emeralds almost always contain some iron, which diminishes the brightness seen. Such a test should not be relied on alone but should be combined with examination under a microscope.

Igmerald

In 1911 work on the synthesis of emerald began at the German firm of IG Farbenindustrie. Early growth runs were hindered by multinucleation, in which many tiny crystals formed, none being sufficiently large to work into desired shapes. Once the growth method was settled, crystals large enough to be faceted as gemstones were grown, but few if any reached the market: many were used for presentations. A successor to the product, given the name of 'Igmerald' (from the firm's name), was grown in the 1950s by Zerfass: we shall look at this emerald below. The Igmerald shows striae parallel to the basal plane of the crystal, and pleochroism is weak compared to that shown by natural emerald. Earlier synthetic emeralds showed strong pleochroism. Characteristic absorption bands are seen, notably at 606 and 594 nm, in addition to those normally shown by emerald. The SG is in the range 2.497–2.702. Inside the stones wisp-like inclusions with tiny bubbles in each separate liquid patch can be seen: the inclusions are grouped in lines resembling swarms, which cross the stone in slightly curved directions.

Nacken emerald

Another very rare emerald (like the Igmerald, keenly sought by collectors of synthetic gemstones) is the material made by Nacken, also in Germany, in the 1920s. Professor Richard Nacken's emerald was long believed to be a hydrothermal product: this arose from a misunderstanding due to Nacken's major work on the hydrothermal synthesis of quartz. A more recent study of the inclusions in the Nacken emerald showed clearly that the growth method involved a flux. Briefly, many stones showed no sign of the presence of water when examined by infra-red spectroscopy: while natural and hydrothermally grown emeralds contain water, those grown by the flux-melt method do not. Some stones were grown on a seed of natural beryl which did contain water, thus explaining an apparent anomaly in which some specimens show the appropriate infra-red spectrum. Inside the crystals (cut stones seem to be very rare), inclusions of the beryllium silicate phenakite can be seen together with twisted veils of flux and two-phase inclusions of a gas bubble in fluid (Figure 9.1).

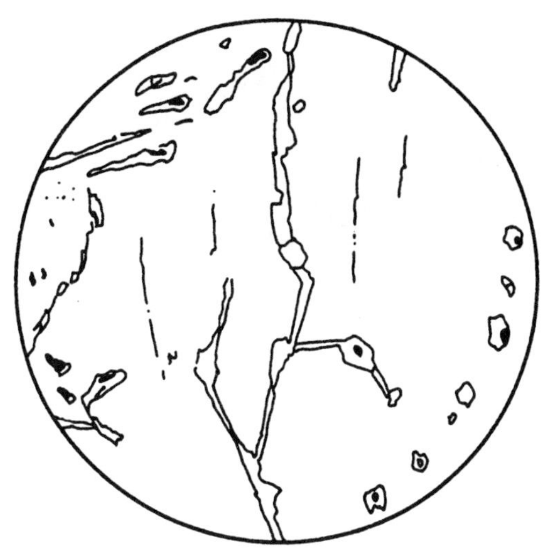

Figure 9.1 Liquid-filled inclusions in early flux-grown emerald

Chatham emerald

Neither the Igmerald nor the Nacken emeralds were commercially significant, the honour of beginning the trade in synthetic emerald belonging to Carroll C. Chatham, who began to grow emerald crystals in San Francisco, California, in the years before the Second World War. After the war he turned to production of emerald crystals for the market, and following his death in 1983 the firm was carried on by his sons John and Tom Chatham. Chatham was a true pioneer, and his product of high quality (and not too difficult to test!). The markets, always conservative, did not take to the stones at first and there were troubles over what they should be called, a truce with the Federal Trade Commission only arriving when the name 'Chatham Created Emerald' was allowed in 1983. As Chatham said at the time, publicity over the name was the best advertising he could have had!

Chatham Created Gems Inc. grows emerald crystals and crystal groups with a growth time said to be measured in months. It is generally thought that a growth period of one year would be needed for a crystal of sufficient size to be fashioned into a stone of several carats. Scientific examination shows that the crystal groups arise from self-nucleation.

The single crystals are grown by the flux-melt process. The infra-red spectrum shows no sign of the presence of water, and Nassau in *Gems Made by Man* (1980) believes that the suggestion sometimes put forward, that natural beryl has been used as feed material cannot be correct. In fact, Nassau suggests that Chatham's process may be similar to the one used by Nacken, since the pattern of inclusions corresponds (as they do also with the processes used by Zerfass, Gilson and Kyocera described below).

The rich green Chatham emeralds have an SG in the range 2.65–2.67 and an RI of 1.560–1.566 with a birefringence of 0.003–0.005. The gemmologist will need to use the microscope for any emerald specimen, natural or suspected synthetic, and the Chatham specimens will be found to contain twisted veils of flux, crystals of phenakite (Figure 9.2), fragments of platinum from the wall of the crucible and two-phase inclusions. There will be no natural inclusions: this, as always, is a very important feature, since natural emeralds are well known to contain fairly profuse inclusions, so much so that the name *jardin* has

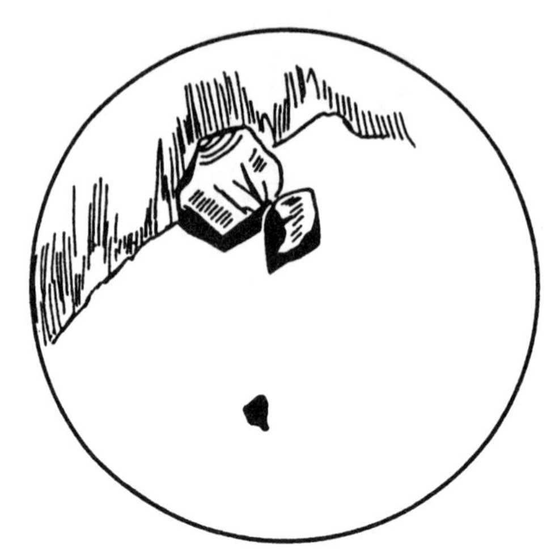

Figure 9.2 Phenakite crystals in a flux-grown emerald

long been used to describe them. The minerals found in emerald can often indicate their place of origin, so that gemmologists are used to looking at them. Crystal inclusions in natural emeralds may be calcite, pyrite, tremolite, actinolite and other species: none of them is found in the synthetic stones, where the only crystal inclusions may be phenakite, which is believed to form when temperatures during the growth process exceed the level needed for emerald growth.

Gilson emerald

In 1950, Pierre Gilson, a producer of ceramic tiles in the Pas-de-Calais, France, took up the synthesis of emerald. Later he also experimented with lapis lazuli, turquoise and coral imitations, as well as a very fine set of opal varieties.

Gilson emeralds grow on a colourless natural beryl seed which, having been coated with emerald, is then removed from the emerald, which in turn is used as seed for the growth of the final product. The rate of growth is reported to be about 1 mm a month so that production of facetable crystals will take a long time, as with the Chatham stones. In both cases the long growth period and its supervision goes a long way towards increasing the final cost to the customer, and synthetic emerald is in fact much more expensive than is sometimes thought.

Textbooks report that at one time Gilson added a trace of iron to the emeralds (the ‘N’ series): there are not many specimens, and the practice did not continue for long. Such stones show an absorption band at 427 nm (together with the normal chromium spectrum). Clusters of emerald crystals have been produced: these are attractive and have a ‘natural’ appearance; however, characteristic swirly patterns on the flat ends of the crystals are not seen on natural emerald crystals, which do not, in any case, grow in this type of cluster.

Faceted Gilson stones were graded by quality, those placed in the highest categories appearing virtually without inclusions: faceted stones up to 18 ct are known. Growth takes place on seeds (otherwise such large crystals could not be obtained): a flux transport

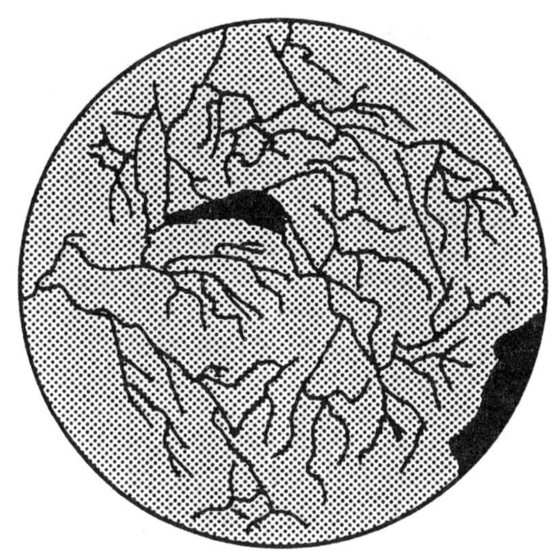

Figure 9.3 Structures reminiscent of cigarette smoke in a still room are diagnostic for emerald grown by the flux-melt method

system is believed to be used. Reports state that the seeds are mounted on a noble metal frame and that the growth rate is approximately 1 mm a month with nine months as an average growth time. Each seed was said to produce about 200 ct of emerald.

A report in 1975 suggested that Chatham and Gilson emeralds could be tested by immersing them in the clear colourless liquid benzyl benzoate. The facet angles were said to light up as the microscope focus was raised. Synthetic emeralds were also said to float (and natural emeralds sink) in a liquid of bromoform diluted with xylol, the correct mixture allowing a specimen of rock crystal to rise slowly to the surface on immersion. Since this test is easily misinterpreted and the constants of both natural and synthetic stones vary quite a lot, the test is not recommended: in any case the use of bromoform is now strongly discouraged on medical grounds.

Chatham and Gilson stones are much better examined under the microscope when the absence of natural mineral inclusions will at once be apparent and the presence of flux particles forming characteristic twisted veils or smoke-like patterns should be noted (Figure 9.3). In the high-quality faceted stones such inclusions that have crept in are often placed near the girdle and can be hidden by the setting. Some specialist manufacturers have set the crystal groups with great success.

These two products still appear on the market although the Gilson process has been sold to the Nakazumi Earth Crystals Corporation of Japan. Today some of the traditional tests may not have the sure results that they once had: emeralds may not always show so bright a red through the Chelsea filter nor be so transparent to short-wave UV. We have already mentioned but should emphasize once more that red fluorescence under LWUV is not common.

Zerfass emerald

The emerald made by Zerfass in Germany during the 1950s (specimens are rare) is believed to be the descendant of the Igmerald described above. The specimens contain profuse inclusions with an overall subhexagonal pattern with pronounced twisting, and it is clear that they are flux particles. The RI is found to be lower (at 1.555–1.561, DR 0.006) than that of benzyl benzoate (RI 1.57), a liquid often used for immersion of beryl since the closeness of the RI values makes examination of specimen interiors easier. This is a fairly low value for natural emerald. The Zerfass emerald gives a weak red fluorescence, and has an SG of 2.66. At least one crystal seen by the author showed a flattish hexagonal prism with tiny emerald crystals protruding from the side faces – very attractive. Faceted Zerfass emeralds are a fine green.

Lennix emerald

An addition to the flux-grown emerald family arrived in 1966, when the first crystals of what was to become the Lennix emerald were grown by Mr L. Lens of Cannes, France. Small emerald clusters were produced first, but by the early 1980s crystals large enough to be faceted were being made. The grower has stated that the flux-melt process is used with crystallization occurring at atmospheric pressure and at temperatures in the region of 1000°C. In a report published in the fall 1987 issue of *Gems & Gemology*, GIA staff examined eight crystals of tabular hexagonal habit, and one faceted stone.

In the crystals the basal pinacoid was the predominating form and the largest specimen examined measured 13.9 × 9.6 × 3.0 mm. The faceted stone weighed 1.30 ct. The colour was dark green and homogeneous, though with magnification an intenser colour could be seen in areas parallel to the *c*-axis, the vertical axis of the crystal.

Some stones examined were fairly clear while others were heavily included – a picture characteristic of flux-melt gemstones in general. Using standard gemmological tests the RI was found to be within the range 1.556–1.568 with a birefringence of 0.003. The RI varies somewhat with colour, higher readings being obtained from the darker-green areas. While the lower values might very well suggest a flux-grown stone, gemmologists will not rely on RI alone. Dichroism does not differ from that shown by natural emerald, and the stones give a bright red colour when seen through the Chelsea filter. The SG falls in the range 2.65–2.66, again not impossible for natural emerald.

The hand spectroscope shows the expected chromium absorption spectrum, and specimens give a bright red fluorescence under LWUV with a weaker red under SWUV. Some chromium-rich natural emeralds will behave in a similar way (this is not very common, however) so that the response of the Lennix emerald cannot be taken as diagnostic. A test using cathodoluminescence showed that some specimens of the Lennix emerald gave a purple or bright violet–blue response, and this has so far not been reported for natural or for any other synthetic emerald.

As always, the pattern of inclusions is the only way in which the Lennix emerald can be distinguished from natural material. GIA found a number of features: opaque tube-like inclusions preferentially aligned parallel to the *c*-axis and clusters of inclusions along the borders of successive growth zones following the edges of the basal pinacoid. Thin prismatic crystals of phenakite and beryl could be seen as well as flux-lined healed fractures of the familiar wispy veil pattern. There were also two-phase inclusions along the edges of the basal pinacoid and occurring sometimes parallel to the *c*-axis. The rather profuse inclusions may superficially resemble the *jardin* of a natural emerald but otherwise the Lennix product can be identified if they are carefully examined. Some stones have also been found to contain opaque black material, probably from a molybdate flux.

Inamori emerald – Seiko emerald

Another successful flux-grown emerald has been synthesized by the Kyoto Ceramics firm (Kyocera), and certainly for a long time stones were only available once set in the range of jewellery distributed through the Kyoto retail outlet. The name Crescent Vert was used for the emerald: in the United States the emerald was called Inamori Created Emerald. The Seiko emerald, another Japanese product, contains planes of radiating phenakite crystals set in groups between growth layers and also single crystals of phenakite. Colour zoning can also be seen with green and colourless layers: near-rectangular scraps of flux have been reported to lie in one general direction in a plane between the colour zones. The RI is reported to be 1.560–1.564 and the SG 2.65.

Russian synthetic emerald

In the early 1980s, flux-grown emeralds from Russia were publicized, and specimens have continued to enter the market. One of the earliest samples, reported in the summer 1985 issue of *Gems & Gemology*, was a cluster of self-nucleated hexagonal prisms, similar to those grown by Chatham and Gilson, the crystals showing the same pinacoidal terminations. In this example they radiated from a crust of polycrystalline material and measured up to 3 cm in length and 4.2 cm in diameter. Faceted stones had appeared in some quantities on some of the world gem markets by the time the paper had been published.

In general, the Russian flux-grown emeralds were similar in properties to other emeralds made by the same method: the manufacture took place at the Geological Institute of Akademgorod, Novosibirsk. A lead vanadate had been used as the flux in place of the lead molybdate used by Chatham and Gilson, and a natural beryl had been used as the nutrient.

The GIA tested a crystal cluster and 18 faceted stones, finding that they could all be distinguished from natural stones by their SG and RI, which were 2.65 and 1.559–1.563, respectively. The birefringence was found to be 0.004. While these properties are low for emerald in general, they are insufficient to identify the Russian stones as undoubtedly synthetic.

The microscope has to be used, and when the crystal cluster was examined its surface showed three distinct solid phases as well as the emerald itself. The phases were a near-colourless transparent material that was found to be phenakite, groupings and single crystals of synthetic alexandrite and silvery metallic platelets of a platinum group member from the crucible liner.

Flux inclusions were observed, and took two distinct forms: one was in the shape of secondary healed fractures while the others occurred as primary void fillings. Some of the inclusions contained two phases with a glassy bubble and others, noted in the faceted stones, showed fingerprint-like patterns. The flux inclusions show that the emerald is synthetic. Reflecting greyish platelets of platinum have also been observed.

Hydrothermal growth of emerald

Growth of emerald by the hydrothermal method is less common than by flux growth, though the two methods and their products share some similarities and have sometimes been confused. The nature of the apparatus has already been mentioned, but the chief difference between the two methods is the use of a sealed pressure vessel, the autoclave,

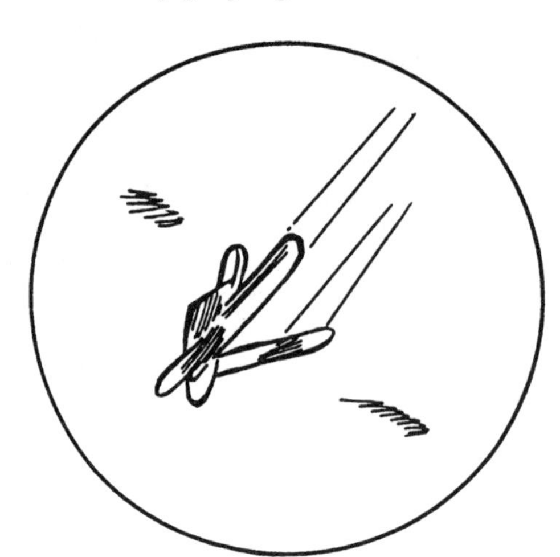

Figure 9.4 Growth tubes emanating from phenakite crystals in hydrothermally grown emerald

in hydrothermal growth. Like most flux-grown emeralds, hydrothermal specimens are grown on prepared seeds under typical conditions of 500–600°C and pressures of 700–1400 bar. The growth rate is in the region of 0.3 mm per day.

Linde emerald–Regency emerald

Hydrothermal emeralds were grown by the Linde Division of Union Carbide Corporation from 1965 to 1970, the author visiting the facility during that time and examining a number of crystals. During the years that Linde grew emerald crystals the precise details of the growth process changed, with the later method giving growth rates up to 0.8 mm per day. The Linde emerald production reached about 200 000 ct a year in the years 1969–70, and during that time the company tried to sell the stones by setting them in their own manufactured jewellery – the Quintessa line. By 1970 too large a stockpile had accumulated and the process was sold to Vacuum Ventures Inc. of New Jersey, who produced the Regency Created Emerald using the same or similar growth methods.

The Linde emeralds have an SG of 2.67–2.69 (well within the natural emerald range) and an RI of 1.566–1.578 with a DR of 0.005–0.006. Features inside the stones are much harder to see than in the flux-grown stones, consisting only of a few phenakite crystals and very fine two-phase inclusions. Some hydrothermal stones give a red flash in strong white light and show tapering growth tubes emanating from phenakite crystals (Figure 9.4). Tiny white 'breadcrumbs' of unknown origin may also be found, and arrowhead-like markings or chevrons may sometimes be seen. No natural mineral inclusions and no liquid inclusions of the type found in the natural stones can be seen.

Biron emerald

In 1977 a Western Australian manufacturer began work on the hydrothermal synthesis of emerald, the stones being sold under the trade name Biron. A report in the fall 1985 issue of *Gems & Gemology* describes the Biron stones, of which 150 faceted examples and 2 rough crystals were examined. The colour varied from green to slightly bluish green with moderate to vivid saturation. Faceted stones were in the main notably transparent with apparently inclusion-free areas. Other stones showed inclusions quite clearly. Dichroism was distinct with a green–blue colour parallel to the *c*-axis and a yellowish green in the direction perpendicular to the *c*-axis. This effect does not distinguish the Biron emerald from its natural counterparts. The absorption spectrum is not remarkable in the visible region, and in the infra-red shows the presence of water, thus proving hydrothermal (or natural) origin.

Biron emeralds have an SG of 2.68–2.71 and an RI of 1.569–1.573 with a DR of 0.004–0.005. These figures are slightly below those shown by many but not all natural emeralds. The stones are inert to both types of UV radiation: iron was not found to be present (the usual 'poisoner' of luminescence when a specimen contains chromium); it is believed that a high vanadium content is the cause of the lack of the expected response. Since many natural emeralds do not respond to UV this is not a very useful test.

As usual the microscope provides the only serious gemmological tool. Some types of inclusion unique to the Biron emerald have been observed while others resemble inclusions found in the emeralds of other growers. Two-phase inclusions of a fluid and a gas bubble form fingerprint-like shapes and curved veils: they have also been seen as large irregular voids containing one or more gas bubbles, and they can also be found in the tapered part of spicules which resemble nail-heads – these tapered formations seem only to occur in hydrothermal emeralds.

In flux-grown crystals, growth defects resembling fingerprints and veils are healed fractures: in hydrothermal emeralds the fingerprints and veils consist of many small two-phase inclusions concentrated at curved and flat interfaces; similar-appearing flux inclusions are solid. Some of the fingerprints and veils in the Biron emeralds dangerously resemble similar structures in natural specimens. The familiar nail-head spicules are cone-shaped voids containing a fluid and a gas bubble: the head of the 'nail' is a single crystal of phenakite or a group of phenakite crystals. The spicules frequently occur in groups which are arranged parallel to the *c*-axis.

Gold-like flakes seen inside the Biron emeralds were found on examination to be that metal, which must have come from the lining of the autoclave in which the emeralds were grown. The gold crystals take a number of different forms, sometimes appearing as angular grains and at others as thin flat plates. Careful examination of the phenakite crystals shows them to be well-formed and up to 0.3 mm in length. It is worthwhile for the gemmologist to become familiar with them since they should not be confused with crystals of natural inclusions such as calcite, which shows recognizable rhombs rather than prismatic forms. A test for those with some experience would be to examine a sufficiently large inclusion between crossed polars: calcite will show brighter interference colours than phenakite. Calcite and dolomite (also forming rhombs in natural emeralds) do not occcur in the same concentrations as the phenakite crystals in the Biron emerald.

Different types of growth features and colour zoning can be seen in the Biron emeralds. While not always easy to see, a 'Venetian blind' effect is sometimes present, and zoning takes several different patterns: again, such effects are not found in natural emeralds. Nor are tiny whitish particles which have been called 'comet tails': seen in some synthetic rubies, they are best viewed with fibre-optic illumination. Some specimens contained seed plates, near-colourless and flanked by planes of gold inclusions.

A low RI and birefringence, characteristic inclusions and, for mineral chemists, the presence of chlorine, which has not been reported from natural or from flux-grown emeralds, serve to indicate the Biron hydrothermal emerald.

The Biron emerald, the Pool emerald and the Kimberley emerald are all of Australian origin and may be the same product. Certainly, emeralds offered by the Emerald Pool Mining Company (Pty) Ltd of Perth, Western Australia, under the name of Pool emerald, turned out to be Biron-type hydrothermal emerald. Pool claimed that their material was 'recrystallized' natural emerald, but in 1988 the Biron trade name was reinstated after being dropped.

Russian hydrothermal emerald

A hydrothermal emerald has been grown by the Laboratory for Hydrothermal Growth at the Institute of Geology and Geophysics, Siberian branch of the Russian Academy of Sciences, Novosibirsk. Compared to other hydrothermal emeralds, the GIA found that they contained different inclusion features although specimens were detectable as synthetic. The emeralds are described in the spring 1996 issue of *Gems & Gemology*.

The RI was in the range 1.572–1.584 with a birefringence of 0.006–0.007: the figures are consistent with those published for previously examined Russian hydrothermal emeralds. They are higher than those for the Biron emerald. The SG was 2.67–2.73, overlapping with natural emeralds and with other synthetic products. None of the eight faceted stones tested showed any response to either type of UV radiation. A weak red glow was seen through the Chelsea filter when the specimen was held at a low angle to the source of illumination. Nothing unusual could be seen with the dichroscope or spectroscope.

Inside the stones none of the chevron-shaped growth zoning characteristic of many hydrothermal emeralds could be seen. Tiny reddish-brown particles were observed to be arranged in dense clouds, but their nature has not yet been determined. Under fibre-optic illumination clouds and layers of minute white particles could be observed in all the stones examined: dense and easy to see in some of the specimens, they were elusive in some others. One stone contained a phenakite crystal and another two opaque black hexagonal plates showing a silvery grey in reflected light. Infra-red spectroscopy indicated the expected presence of water. The Russian emeralds can best be identified by the reddish and whitish particulate inclusions, too small to have been identified as yet.

Russian hydrothermal emeralds have been found to contain traces of nickel and copper, which have not been reported from natural stones, traces of water detectable by infra-red spectroscopy (water is also present in natural emerald) and sets of parallel lines with a step-like formation, sometimes connected to colour zoning. Growth takes place on seeds, and rates are reported to be fast. This relatively fast growth rate gives rise to the step-like structures which lie parallel to the seed surface. A report in the spring 1996 issue of *Gems & Gemology* describes fragments or slices of what must have been larger crystals: two of the pieces contained colourless or slightly greenish residual portions of the seed. Faceted stones were also examined.

Parallel growth planes forming an angle of approximately 45° to the optic axis were observed: seeds were cut parallel to a face of the second-order hexagonal dipyramid (a note for crystallographic readers) with the aim of avoiding characteristic growth patterns seen in previous Russian hydrothermal emeralds. Growth is said to have taken place in steel autoclaves with no noble metal inserts.

Figure 9.5 Emerald overgrowth on pre-cut seed

Lechleitner emeralds

In the early 1960s, Johann Lechleitner of Innsbruck, Austria, grew a thin layer of hydrothermal emerald on to faceted natural beryl seeds. The stones were then polished (sometimes leaving back facets in their original state). They were marketed by the Linde Company in the United States and by an Austrian firm: trade names used, at least for a time, include *Symerald* and *Emerita*.

The colour of the Lechleitner stones is very attractive even though the emerald overgrowth is thin. When specimens are immersed a characteristic crazing can easily be seen: this is the best way to identify the stones. Lechleitner has also grown 'full' hydrothermal emeralds.

The Lechleitner overgrowth emeralds show not only the crazy paving markings but also parallel lines along the length of the stone (Figure 9.5). Tiny dust-like crystals of euclase or phenakite occur at the junctions.

Lechleitner has also produced a sandwich emerald in which a seed plate of colourless beryl is coated by emerald by the hydrothermal process. Then the specimen is further coated by hydrothermal growth, this time to give a colourless coating. Closed settings would be needed to conceal the different layers.

It is interesting to find that the fine colour of the Lechleitner emerald overgrowth stones is widely acknowledged. A report in 1981 *Gems & Gemology*, summer 1981 issue, shows that the emerald coating has a very high chromium content: if the same amount of chromium were present in a 'whole' emerald, the colour would be too dark for the stone to be commercially acceptable. Lechleitner is also believed to have made complete emeralds.

The name Emeraldolite was given to a green material consisting of an epitaxial growth of flux-grown emerald on to opaque white beryl. It is not of faceting quality but is suitable for cabochons. The material was manufactured in France: the process of growth is to some extent similar to that used by Lechleitner, except for the use of a flux rather than a hydrothermal growth method. Emeraldolite is opaque, and a coating of only 0.3–1 mm in thickness is used, thus giving an inexpensive product. One difficulty is that growth takes place differentially, so that spheres cannot be made because shapes are asymmetrical. The surface of polished cabochons would give rise to suspicion if they were offered as natural emerald, since they are so irregular and uneven in appearance.

The SG and RI are 1.56 and 2.66, respectively, both figures characteristic of flux-grown emerald. Both core and overgrowth were inert to UV. There was a chromium absorption spectrum, and specimens showed brownish red through the Chelsea filter.

The hardness (which could safely be tested on this material) was found to be approximately 8 on Mohs' scale. The material is also tough, and the overgrowth cannot be separated from the core by breaking. Under magnification the flux overgrowth layer could be seen to show several groups of minute parallel crystal faces. In places where the overgrowth was missing, the white beryl could be seen below. The emerald crystals in the overgrowth could be seen to display the normally expected forms of prisms and basal planes: some bipyramidal faces were also noted, though the 'crazy paving' cracks associated with the Lechleitner emerald overgrowth were not seen. In the synthetic emerald layer fairly large flux inclusions could be seen, with high-relief more or less spherical voids looking very like the large gas bubbles which can be seen in most glasses. Crater-like pitting on the surface arises when these bubbles are open.

Aquamarine

The lower commercial value of aquamarine makes synthesis less of an attractive commercial proposition than that of emerald: additionally, the easy availability of imitations made of synthetic blue spinel grown by the very cheap Verneuil method may deter many manufacturers. None the less, aquamarine has been synthesized in Russia by the hydrothermal method, the growth taking place at the Institute of Geology and Geophysics, Siberian branch of the Russian Academy of Sciences, Novosibirsk.

The aquamarines obtain their colour from iron in both ferrous and ferric form. Crystals have an SG of 2.69 and an RI of 1.575–1.583 with a birefringence of 0.008. The visible and UV absorption spectrum shows the presence of iron and nickel, and the infra-red spectrum shows the presence of water, thus proving hydrothermal growth. Under the microscope the near colourless seed and light blue overgrowth show a clear boundary, and very weak colour zoning can be seen in the aquamarine parallel to the boundary with the seed plate.

It is likely that the seed is also hydrothermally grown beryl since its inclusions resemble those in the overgrowth. Both portions display a cellular structure seen best when the specimen is immersed and at about 60× magnification. Groups of small birefringent crystals and some opaque hexagonal plates can be seen: the plates are probably hematite. There are also cavities with multiphase fillings, and two different kinds of feather-like structures, one type nearly flat and the other forming twisted veils of trapped growth solution.

For the gemmologist the easiest way to distinguish the hydrothermal aquamarine from the natural stone is by using the microscope to detect the cellular structures, the groups of doubly refractive crystals, the cavities with multiphase fillings and the different types of feather-like structures. The absorption spectrum of nickel, though not able to be seen with the hand spectroscope, is not so far recorded from natural aquamarine: the nickel comes from the pressure vessel, which does not contain a precious metal lining. Chemical analysis will show magnesium and sodium to be present in amounts which, with a high iron content, show that specimens are synthetic.

While I have seen enough reports of synthetic emerald and aquamarine to know that the gemmologist must learn to use the microscope and interpret what is seen, laboratory workers will more often use the spectroscope to detect elements foreign to the natural material. The spectroscope in this context is not the 'direct vision' instrument which is used only with visible light but a much more complex apparatus that is sometimes known as a spectrometer as it records the position of absorption bands in wave numbers (the reciprocal of wavelength).

In the beryl context the spectrometer looks for such features as the absorption bands in the infra-red which occur when water is present in a crystal. The rings which atoms of beryl form allow a central channel in which water can be present: when seen, these bands rule out flux growth but allow hydrothermal and natural origin. When the spectroscope is used to test anisotropic specimens the crystallographic orientation of the specimen should be ascertained and readings taken in different directions to be sure of a satisfactory result. This is not of course easy when the sample is a faceted gemstone. None the less, when absorption features are seen their strength compared to the strength of corresponding absorptions in natural material is an indication of a possible artificial stone.

While the absorption spectrum of water allows the laboratory to distinguish between natural or hydrothermal beryl on one hand and the flux-grown material on the other, the distinction between natural and hydrothermal is also possible. Most types of hydrothermal emerald show absorption features which will distinguish them from their natural counterparts. Further details can be found in the paper by Stockton in the summer 1987 issue of *Gems & Gemology.*

When green beryl looks like emerald and is found to contain chromium, it can be called emerald by the gemstone and jewellery trade. If it is beryl, looks like emerald but contains no chromium, it must be named green beryl. This causes confusion from time to time, and for years, when as a curator at the British Museum, I used to pass a jeweller's window daily, a fine light sea-green stone was on display as emerald, though it looked much more like the 'old' type of aquamarine. At length patience snapped and I borrowed the stone for testing: sure enough it showed a chromium spectrum, it was beryl and so was quite correctly named. It was a beautiful stone, and I now wish that I had bought it!

Not only chromium can give an emerald-green colour to minerals: when it can be suitably accommodated at the atomic level, vanadium can also. In the 1960s, Taylor at the Crystals Research Company of Melbourne, Australia, grew hydrothermal green vanadium beryl which as faceted stones looks very like emerald.

Crystals were reported at the time to reach 10 ct in size and to yield faceted stones of up to 2 ct. The properties accorded with those shown by natural beryl with an RI of 1.566–1.575 and a birefringence near 0.005. When specimens were immersed in a liquid of similar RI, marked colour banding became visible in the earlier stones, though specimens produced later showed the effect much less strongly. The SG was in the emerald range at 2.68. Specimens were inert to both types of UV and to X-rays, and the body colour was a warm grass green with a trace of yellow, with less blue than in many other emeralds.

Dichroic colours were yellow–green and green, and through the Chelsea filter specimens showed greyish green to dull pink. When a polaroid was used with the filter it was found that the ordinary ray was green and the extraordinary ray pink. The absorption spectrum was normal for emerald and gave the water absorption bands in the infra-red region. Traces of the seed could be seen inside the specimens and were placed at an angle to the *c*-axis. Their thickness was estimated to be approximately 1 mm. Interfaces of seed and overgrowth were visible on immersion, and no solid inclusions could be found.

Before leaving emerald we should remember that there are several types of green composite stones which can give a convincing imitation. Often known as soudé emeralds, they can take various forms but where the crown of the stone is glass the RI will give it away: the absorption spectrum will show woolly bands rather than the sharper ones of emerald, and in any case the stones will lack both natural inclusions and the flux traces of the synthetic stone.

Other colours of beryl

Beryl of other colours than emerald or aquamarine has occasionally been synthesized. The Japanese firm Adachi Shin Industrial Company of Osaka has grown a 'water melon' beryl with a pink core and green skin – reminiscent of tourmaline. The RI was 1.559 and 1.564 and the SG 2.66 – figures well below those of tourmaline. Specimens were reported to be grown by a method in which fluorine and oxygen react at higher temperatures than are usually used in beryl growth with crystalline or amorphous beryllium oxide, silica and alumina. Dopants are added to give the colour (possibly chromium for the green and manganese for the red), and after heating the melt migrates to a cooler zone of the apparatus amd then on to seeds. Hydrothermally grown red beryl, doped with cobalt, has been made experimentally.

Summary

Looking back over the synthetic emeralds manufactured now for more than 100 years, several points are clear: growth by simple and cheap methods is not possible for emerald at present, and the other varieties of beryl are already easily imitated by glass and by synthetic spinel. Synthetic emeralds are still not cheap, for reasons we have discussed above, but are plentiful enough for any jeweller not to have to wait too long before meeting one.

While the Chelsea filter still provides a useful means of enhancing suspicion that should always be in the mind, those dealing with emeralds should always have a microscope on hand. While gemmological textbooks always advocate the $10\times$ lens, the presence of a microscope, perhaps obtained at some expense, will lead owners to use it first. The lens certainly is useful but not really easy to use. Other gemmological instruments, useful though they are, play little part in the identification of synthetic emerald since its properties almost invariably overlap with those of the natural mineral. It is not possible to describe the different types of synthetic emerald without some citation of growth process features: crystal growth is complicated and its outcome by no means predictable.

As always, the gemmologist should learn to look at the interior of natural emeralds (hoping that they will provide 99 per cent of the emeralds submitted for test!) and then examine known synthetics to see what the differences are. Almost every natural specimen will show recognizable mineral inclusions which are not present in the synthetic emerald, which in its turn will show the presence of flux, of unusual colour banding or of chevron-like markings or spicules.

Gemmologists have to face another problem with emerald: from very early times the colour has been improved by a number of methods, which have recently been joined by fresh ones.

The colour enhancement of emerald

Although a visit to London's Bond Street and Burlington Arcade will quickly convince the visitor that most emeralds are a rich deep green, geologists familiar with emerald mining will know that crystals of a fine colour are rare: examining emerald crystals at mines in Pakistan, I found that most were a pale green, not too pale for jewellery use but not the rich green of the finest stones; these did turn up but were in the minority.

Most emeralds are known to contain internal furniture: inclusions to the mineralogist or gemmologist and *jardin* to some of the trade, they are known to the public as flaws. A large emerald of fine colour but containing few inclusions would look too glassy to be really attractive: inclusions scatter light and make the stone appear a softer green.

The colour of pale emeralds has been deepened over the centuries by a variety of treatments, some working better than others. Stones with too many inclusions can also be treated to diminish their effect – that is, to reduce light scattering. The history of gemstone treatments is admirably described by Nassau in *Gemstone Enhancement* (second edition, 1994), and the reader is particularly recommended to consult this book for accounts of treatments in history as well as for current practices.

We are concerned with identifying enhancement when it has been carried out: for this we need to know something about the processes used. The commonest treatment of emerald is oiling: immersion of the stone in an oil which fills surface-reaching fractures, making them less obvious to the viewer. The practice goes back at least to classical times and suggests that emerald crystals then available were mostly pale (this can be borne out by an examination of present-day Egyptian deposits). The oil can be chosen from an extensive list, but in general the lighter oils work best, the heavier ones tending to ooze from cracks when the stone is heated. When the oil used has an RI similar to that of emerald, treatment is particularly effective: Canada balsam has an RI of 1.53, close to the emerald range of 1.57–1.59 and has been successfully used, though probably the commonest vehicle is light sewing-machine-type oil, which is universally available. Nassau (1994) cites a paper by Ringsrud in the fall 1983 issue of *Gems & Gemology* in which oiling of Colombian emeralds is described: the report states that after faceting the stones are first cleaned by boiling several times with methyl or ethyl alcohol, followed by the application of aqua regia in a closed container with or without ultrasonic cleaning (this will very often fracture emerald). The acid is used to remove particles of tin oxide and chromic oxide which have entered the fractures during the faceting process – its use is very dangerous and storing the mixture in closed containers particularly so. Do not try this at home.

Sometimes emerald crystals are oiled when they are to be put on sale. The acid treatment reacts with the oil to give permanent brown stains inside the stone, which may then present a very natural appearance. Otherwise, after acid treatment, faceted stones are boiled or slightly heated in alcohol, ethyl alcohol, acetone or paint thinner. Some believe, however, that boiling in water is to be preferred. Oiling now takes place, Ringsrud stating that in Bogotá cedarwood oil and Canada balsam are the favourite vehicles. The former is much less expensive and used the most. The oils are heated to facilitate introduction into the fractures, and sometimes this is done in a vacuum. Stones are then polished and ready for sale.

Treatment of emerald in this and other ways is virtually universal, and it is thought remarkable if a stone has not been improved. Some high-clarity Colombian emeralds are not treated, however, since their appearance could not be improved. When I was working in Pakistan in the late 1980s, stones were not being oiled and very fine qualities were being mined.

The gemmologist's problem begins when confronted by a single oiled specimen or by a parcel of loose oiled stones. The second scenario is much the easier to deal with since the oily smell emanating from the opened parcel is unmistakable: examining the contents of the parcel will immediately coat the fingers with an oily substance, again, unmistakable. For the single faceted emerald the gemmologist will certainly need the suspicious mind before anything else: anyone routinely examining emeralds should look for signs of oiling.

While some fracture-filling substances will give a dull yellow fluorescence under LWUV, Ringsrud did not find this with cedarwood oil as used in Bogotà. On the other hand, the Canada balsam did fluoresce, and thus can easily be detected when present. Side lighting from a fibre-optic source may show the filled fractures, which will appear dull. They should not be mistaken for natural liquid-filled inclusions, so care needs to be taken to ensure that the suspected filled area reaches the surface. If there is a marked difference between the RI of the oil and the emerald, spectrum colours may be seen in the filled fracture. Use of the thermal reaction tester may cause a bead of oil to appear on the surface: Ringsrud mentions that advice from a Bogotá dealer was to examine the paper of the stone parcel for signs of oiliness.

Canada balsam and cedarwood oil are colourless but coloured oils are also used to fill fractures in emeralds, though in 1983 Ringsrud was told that this was not a common practice. Under magnification, areas containing a coloured oil would show a concentration of the colour which in any case would not be exactly the same as that of the host. If a suspected stone is placed on a translucent white plastic background over an intense source of light, colour concentration will easily be seen in the diffused light. It is important that when areas of deeper colour are seen they should be examined from all possible angles, since unfilled fractures will still collect and reflect green light from many parts of the stone.

A more recent development in emerald enhancement is the filling of surface-reaching fractures by glass or epoxy resins: Opticon is probably the best known of the epoxies, so that any kind of filling is sometimes called Opticon whatever its true nature. By the 1990s a number of firms were offering an emerald-filling service, increasing problems for the gemmologist. As well as identifying emeralds with fractures filled in this way, the gemmologist has to try to ascertain the durability of the fillings – whether they will degrade, fragment or even leave the specimen.

The full name of Opticon is Opticon resin No. 224: developed by Hughes Associates of Excelsior, Minnesota, USA, it is reported to be used in a number of emerald-mining areas, including Santa Terezinha, Goiás, Brazil, where it is applied to 5000–6000 stones per month, according to an authoritative report in the summer 1991 issue of *Gems & Gemology*. Most of the treated emeralds are in the 0.5–2 ct range. After cutting, treatment is said to begin with cleaning by soaking in dilute hydrochloric acid and rinsing in water. The acid is weak, and lemon juice is sometimes used as a substitute. Opticon may also be applied to the rough material, and will have to be removed in the cleaning process.

Opticon is then applied to the stones, which are placed in heat-resistant beakers: low temperatures are applied (95°C has been quoted) with a view to reducing viscosity and making penetration into the fractures easier. After heating, the stones are allowed to cool to room temperature while still immersed in Opticon. They are then removed from the beakers, and a hardening agent applied to the surface. Left there for 10 minutes, the hardener allows the top part of the filler to set while the remainder of the Opticon is still in liquid form deeper inside the fracture. After 10 minutes the hardener can be removed only by repolishing. Up to 25–30 stones can be treated with the hardener at one time, and after this process is complete the stones are washed in a solution of water and baby shampoo, then rinsed in water.

Stones are then examined for any damage that may have occurred during treatment: chipped edges are a common result, and such stones have to be recut and treated again. There are many variations of this process, some involving treatment in a vacuum and heating of the filler by a radio-frequency thermal wave transmitter.

While the Opticon so far described is colourless there is a green variety which has also been used to fill fractures in emerald: it is not the only coloured fracture filler. It has been reported that stones with fillings of the coloured Opticon break more easily, but this is not of course a useful test! Experiments with the coloured material show orange and blue dispersion flashes from the filled fractures.

Opticon has an RI of 1.545, higher than those of Canada balsam and cedarwood oil but lower than that of emerald. Experiments conducted by the GIA show that emeralds so heavily included that they appeared whitish were changed to a green after fracture filling with Opticon. Reduction in light scattering from profuse inclusions also produces an improvement in colour.

Testing Opticon-treated emeralds should begin with visual observation and we have already mentioned the blue and orange dispersion flashes which cannot be confused with any other effect. Where there are numerous filled fractures the stone will appear less transparent or give a distorted effect which is not a feature of untreated emeralds. The effect is reminiscent of heat shimmer. Some filled fractures may give a weak chalky white to blue fluorescence under LWUV (Opticon alone shows this effect); there is no response to SWUV.

The best test is magnification, and under the microscope filled fractures show very low relief. Unfilled fractures containing air will have high relief and be easy to see. Gemmologists should examine the surface to see where fractures reach it.

The lighting should be arranged in such a way that only the surface is seen by reflected light – a fibre-optic source is ideal for this purpose. Look for fine lines outlining the top of the fracture. Under dark-field illumination arrange the specimen so that light is reflected from the surface you are examining and keep the stone low in the light bowl. When the top of a fracture is seen, the portion of the stone lying beneath it should be carefully examined.

The orange and blue dispersion flashes are distinctive and resemble those seen in fracture-filled diamonds. The GIA noted that when the filled fractures were examined nearly edge-on some showed a slightly orange–yellow colour. Sometimes the whole or a large part of a fracture appeared to take on this colour: on tilting the stone this colour changed to blue. In other examples the orange flash did not turn to blue but disappeared as the stone was tilted and reappeared when the stone was returned to its previous position. Flashes in Opticon-treated near-colourless beryls showed the orange flash alone more often than the blue flash. Coloured flashes have not been reported from stones treated with Canada balsam or with cedarwood oil.

Bubbles surrounded by cloudy areas can be found in the fillings: some of the bubbles show interference (rainbow-like) colours when examined by overhead lighting. They may occur singly or in groups. Some of the fractures show a flow structure best seen in dark-field lighting. The thermal reaction tester may cause some of the filler to sweat out of the stone, though this is less likely to happen when the fracture has been sealed with a hardener. Examination by microscope with the specimen immersed is useful, though care needs to be taken that the liquid does not cause the filler to dissolve.

How long the Opticon filling will last is not yet known: a test conducted by the GIA on an Opticon-treated specimen showed that this filling was more durable than fillings using cedarwood oil or Canada balsam. After fairly strong and repeated ultrasonic cleaning some filled fractures had lost some of their substance, and since such treatment could be expected to be undergone by many filled emeralds in the course of their life in jewellery and because the loss of material could be seen with the lens since previously invisible fractures were now visible, it is possible that charges of stone switching might be made when a stone is returned from the jeweller who may have had it cleaned in this way. In practice, none of the beryl gemstones should be placed in ultrasonic cleaners since they are notably brittle.

Steam cleaning was found to remove more material from the filling than ultrasonic cleaning, and stones whose Opticon-filled fractures had been sealed lost some material. Some Opticon altered from a pale to a medium yellow after about two hours in a test involving the retipping of prongs, a common jewellery repair practice. In some cases Opticon flowed from the stone: it is not recommended that retipping is undertaken with any gemstone in place, however. When the surface of an Opticon-filled stone was polished with a metal polish, the Opticon reaching the surface was abraded by the process.

A low-polymer epoxy resin known as palm oil has also been used to fill fractures in emerald imported into Japan from Colombia, according to the GIA. It has an RI of 1.57 (in the beryl range) and shows weak bluish-green to orange-red dispersion flashes. Trapped bubbles are also found. Other fillers reported in emerald include a cyano-acrylic, giving a white brushmark structure in fractures, and a hardened epoxy resin with no dispersion colours and very few bubbles.

The colour enhancement of beryl

Apart from emerald the varieties of beryl are not often colour enhanced. Many aquamarine crystals when mined are a sea green and very attractive to many: the rather metallic blue of aquamarine found in modern jewellery results from heat treatment but this is so common that disclosure is not expected nor is there an easy test for its diagnosis. A number of heat and irradiation treatments have been carried out experimentally on the other colours of beryl but none is believed to be in commercial use.

There is one example, however, which illustrates more than one aspect of treatment and of the gemstone trade. In about 1917 a deep-blue beryl with a colour resembling a blue sapphire rather than aquamarine was found at the Maxixe mine in Brazil. The colour was very attractive but faded after a time: the colour could in theory be restored by irradiation, but there is little commercial sense in doing so if the stones already have a reputation for colour instability. So until the 1970s Maxixe blue beryl was a gemmological curiosity.

In the early 1970s a fine deep-blue stone was brought to my gemmology class at London Guildhall University. Encouraging students always to use the spectroscope first on any strongly coloured stone rather than reaching automatically and unimaginatively for the refractometer, I looked first and saw that it could be neither sapphire nor a dark-blue synthetic spinel, since the absorption spectrum, with bands in the red and the yellow, was inconsistent with either. The thought 'Maxixe beryl' crossed my mind: the ordinary ray was a darker blue than the extraordinary ray (the reverse effect from aquamarine), and this, with the beryl RI, confirmed my suspicions that here was a Maxixe beryl.

At that time a blue beryl with the trade name 'Halbanita' was being mentioned in the gemstone trade and advertised in gem and mineral magazines. Once the literature became specific it was clear that the stone we had examined was the same material, and its identity was established at the laboratory of the Deutsche Gemmologische Gesellschaft on whose council I then was. By that time the London gem trade was quite excited about the new blue stone and several dealers informed me that they had good stocks – the price, I still remember, was £85 per carat for 1–2 ct stones.

Slowly the rumour spread – the dark-blue beryl colour was unstable. The rumour must have begun in gemmological circles since gemstone traders do not have time to test their stones to see if the colour fades. Naturally, as with all rumours, there was a good deal of exaggeration. The stones 'faded in light', 'could not be worn in jewellery' and so on. The only redeeming feature was that the London trade congratulated themselves on seeing through the stones at once and on not having wasted their money. This was a relief anyway, and the phrase '£85 per carat' clearly must have been a voice from a dream!

For years a distinction was made between Maxixe beryls as first mined and Maxixe-type beryls, looking the same but beginning life as colourless or pale-pink beryl, the deep blue being created by natural irradiation in the first case and man-made irradiation in the second. While the cause of the deep-blue colour is established, a recent paper by Nassau in the *Journal of Gemmology* for April 1996 shows that tests originally used to check the rate of fading involved exposure to greater temperatures than would be expected during normal wear: further reports on the colour stability were in preparation at the time of writing in June 1996.

In the same paper, Nassau emphasizes that aquamarines do not fade on exposure to light but that any yellow component in greenish aquamarines will be lost slowly at 100°C and more rapidly at higher temperatures. The same is true of golden beryls. The temperature quoted is unlikely to be encountered during wear but might occur during repairs to jewellery.

Reports of interesting and unusual examples from the literature

Items in this section have been chosen to illustrate points made in the chapter and to bring one-off items to your notice.

Glass with a colour resembling that of top-quality emerald but with radioactivity up to 12 times background levels is reported in the *Journal of Gemmology*, Vol. 23(2) (1992). The specimens had an RI of 1.635 and an SG of 3.75–3.76.

Gemmologie, Vol. 45(1), includes notes on two emerald imitations. The first consisted of a faceted natural emerald, probably Colombian, overgrown by a layer about 2.5 mm thick of green artificial glass. The origin of the specimen was thought to be Russia and the weight was 5.62 ct. It gave an RI of 1.570–1.578: one of the back facets gave a single RI of 1.547, which could arouse suspicion if anyone thought about testing a back facet in the first place.

The second stone was a crystal whose core was made from a green artificial resin. The specimen gave strong anomalous DR between crossed polars, and also showed bubbles in the core.

On 28 July 1990, Johann Lechleitner was 70 years old and the *Australian Gemmologist*, Vol. 17(12) (1991), summarized his productions to date. *Type A stones* comprise emeralds manufactured by flux growth on to seed plates of natural beryl, the plates cut in a direction perpendicular to the *c*-axis: these emeralds were grown between 1958 and 1959 and sent only to gemmological laboratories rather than being placed on the gem market. *Type B stones* (1959–72), which were sold commercially, had cores of colourless or slightly greenish natural beryl with a hydrothermally grown overgrowth of thin layers of emerald. Both Fe^{2+} and Fe^{3+} ions are detectable (from the colourless core) with the spectroscope, along with chromium, which gives the colour in the emerald layer. Fissures and cracks can be seen in the overgrowth. *Type C stones* used natural beryl seeds cut obliquely to the *c*-axis and the seed: they were reported to be grown in a single run, and the seed shows dark-green emerald layers on both sides of a colourless centre. Stones did not enter the market and were grown from 1962 to 1963. *Type D stones* are complete and are grown hydrothermally, using seed plates from the hydrothermal emerald layers from the type C material. They have been grown since 1964. Examined with the microscope when immersed, specimens are seen to have a layered structure. *Type E stones* were grown for research purposes only, while *type F stones* were grown between about 1972 and the 1980s, and are flux-grown emeralds. Only types B, D and F emeralds are commercially available.

Stones can be separated from natural emerald by their inclusions with growth features, fissures and cracks as well as characteristic signs of flux and hydrothermal growth.

Biron, manufacturers of hydrothermal emerald, has also produced a pink beryl which is described in the *Australian Gemmologist*, Vol. 17(6) (1990). The Australian firm's product shows constants normal for pink beryl but contains no natural inclusions and gives a moderate apricot fluorescence under LWUV.

When fractures in emeralds are oiled they may show a yellow fluorescence under UV radiation. If whitish dendritic deposits are noticed in the fractures it indicates either subsequent ultrasonic cleaning or exposure to high temperatures at the time of fashioning into jewellery.

Aquamarine with an iridescent coating is reported in the spring 1984 issue of *Gems & Gemology*. It has been reported on golden beryl and on emerald. The coating can be removed with light polishing. *On emeralds at least, the coated surface appeared to have been poorly polished.*

An emerald, perhaps of Russian origin, was described in the January 1996 issue of the *Journal of Gemmology*. The stone had been filled, probably by a resin other than Opticon 224 and perhaps Araldite NU471 or Novogen P40. *This belief was due to the very strong colours (orange, blue, yellow and purple) seen (depending on the angle of viewing) under appropriate lighting conditions and on the reaction of a fracture to UV radiation. The fracture fluoresced moderate yellow/white and gave a persistent phosphorescence under SWUV: another fracture fluoresced strongly yellow to white with a very short phosphorescence under SWUV.* A hardener may have been used to seal in the resin since there was no surface reaction to the thermal reaction tester.

In some hydrothermal emeralds examined at the Bahrain Gem and Pearl Testing Laboratory and reported in the April 1995 issue of the *Journal of Gemmology, immersion in benzyl benzoate and examination with the microscope using dark-field lighting conditions showed chevron-like markings and tiny white pinpoint inclusions running in lines throughout the stone, with two-phase inclusions and one example of a spicule or nail-head inclusion: the latter is particularly characteristic of hydrothermal emeralds. Some band-like colour zoning could be seen in one stone which gave a low RI reading of 1.565–1.570.*

Synthetic water-melon-coloured beryl (pink core, green rind) made by the Adachi Shin Industrial Company of Osaka, Japan, was reported in the spring 1986 issue of *Gems & Gemology. Normal testing showed the material to be beryl rather than a tourmaline group mineral: the SG was 2.66 and the RI 1.559–1.564. The core showed strong signs of growth zoning with tiny single-phase inclusions not entering the rind.*

From the company the GIA learnt that a new method of synthesis had been used, in which fluorine and oxygen react at higher temperatures with crystalline or amorphous beryllium oxide, silicon dioxide and aluminium oxide with various dopants. For the water-melon beryl. manganese gives the pink inner and chromium the emerald-green rind. The company reports that crystals up to 1 cm in size have also been grown with brown, reddish brown, pink, colourless, sky blue, yellowish green and emerald green colours.

An emerald set in a ring showed inclusions characteristic of Zambian material and gave the expected absorption spectrum to the GIA, who report it in the spring 1986 issue of

Gems & Gemology. However the RI was found to be 1.48 rather than the usual reading in the 1.57 area. Using reflected light and viewed under the microscope an iridescent coating could be seen: this could be removed with an ink eraser, upon which further testing gave RI values of 1.579 and 1.588.

Emeralds which were in fact rock crystal with a green backing were found by the GIA in a segmented reversible necklace. The heart-shaped green stones gave an RI reading of 1.54, too low for emerald, and were seen to contain two-phase hexagonal inclusions. The stones were in a closed setting, and it was not possible to determine the nature of the backing. The necklace was quite elaborate, and proves the point that even expensive-looking jewellery may not be all that it seems.

A triplet imitating emerald and made from synthetic spinel and glass is featured in the winter 1986 issue of *Gems & Gemology*. The GIA report that this type of triplet was first made in 1951 by Jos Roland of Sannois, France, where the stone was called *soudé sur spinelle*. Stones were made in various colours by sintering coloured glasses to the colourless spinel crown and pavilion. The stone examined by the GIA weighed 11.66 ct and was emerald-cut. Between crossed polars the triplet had a cross-hatched appearance and curled black bands, effects characteristic of synthetic spinel. Interestingly, the RI taken from the table showed a reading of 1.724 (strong) and a weaker one at 1.682.

Between the two readings some other shaded areas could be seen. From the 0.55 mm thick green glass layer an RI of approximately 1.682 was obtained. The hardness of this layer was about 4, a figure shared with many other highly refractive glasses. No absorption bands could be seen with the hand spectroscope. *Under the microscope small flattened, rounded and irregularly shaped bubbles could be seen in the separation plane: these showed best when fibre-optic illumination was used. Looking perpendicularly to the girdle, the thick glass could be seen to show rounded edges and very prominent swirls. Immersion in di-iodomethane showed the composite nature of the specimen very clearly. Under LWUV the crown showed a strong chalky yellowish-white fluorescence when viewed nearly perpendicular to the girdle with the table closest to the UV source; the glass layer was inert and the pavilion gave a strong clear yellow fluorescence with no trace of chalkiness. When the culet was placed close to the source an opposite effect could be seen, with colours reversed.* The GIA believed that this effect was caused by the glass diminishing the amount of radiation reaching those parts of the stone that were not directly facing the source. Under SWUV the stone was virtually inert, and with X-rays a very weak chalky green fluorescence was seen. There was no phosphorescence.

It is not easy to imagine gemmologists being taken in by this ingenious composite but, as always, with an unknown behaving oddly, you have to 'think composite'. Specimens of the same type have been reported in yellow, purple and orange–red colours.

An imitation emerald crystal was made by breaking a pale-green specimen in half across the length, then drilling out the cores of the halves, replacing them with a dyed green epoxy or plastic substance. The two sections were then cemented together. In the summer 1989 issue of *Gems & Gemology* the specimen was reported to contain *numerous gas bubbles which could be seen through any of the long prism faces. The SG was only 2.36 compared to the 2.70 characteristic of natural emeralds.* The crystal showed no pleochroism as emerald would, and the absorption spectrum was typical of a dyed green material.

The Pool hydrothermally grown emerald was the subject of the *International Colored Gemstone Association Alert* No. 18 of 1988. The emeralds were also the subject of a further alert on the basis of their properties very closely approaching those of the Biron emerald, if not identical with them. Biron emeralds are also grown by the hydrothermal process. It is possible that they are the same product.

Synthetic beryl crystals grown in Russia feature in the winter 1988 issue of *Gems & Gemology*. The crystals are unusual for their colours, which include purple (from doping with chromium and manganese), red (from doping with manganese alone), blue (doped with copper) and a rich orange–red (with cobalt). Writing in 1996 it would seem that the products remained experimental only, since none appeared on the market.

Imitation emerald crystals composed primarily of quartz cemented by a green binder are characteristic of material offered to the gemstone trade in Africa. In the summer 1990 issue of *Gems & Gemology* two such crystals are described. The larger weighed 63.35 ct, and both were dark green with rough surfaces partially coated with a light orange–brown staining and with small mica flakes. *The brown staining suggested the presence of limonite. On one area an RI of 1.545–1.553 with a DR of 0.008 was obtained, which indicated quartz. The absorption spectrum showed features consistent with the presence of green dye. Under the microscope the crystals could be seen to be composed of fragments upon which striations were randomly oriented from one piece to another. The green binder contained many gas bubbles and could be easily indented with a needle probe.* The mica flakes were glued on the crystal faces.

An excellent laboratory account of a fracture-filled emerald is given by the GIA in the spring 1990 issue of *Gems & Gemology*. The emerald, weighing 14.84 ct and of fine colour, was found to have surface-reaching fractures filled with a foreign material. Gemmological tests gave readings consistent with emerald, though under LWUV, while the stone itself showed no fluorescence, *some of the surface-reaching fractures showed a very weak, dull chalky yellow response, a weaker effect than that seen in oiled emeralds. When magnified, the stone showed a flash effect, very similar to the effect seen in fracture-filled diamonds. Using dark-field illumination, almost all of the large surface-reaching fractures showed a yellow to orange interference colour: as the stone was moved to a bright background this changed to an intense blue. Tilting the stone backwards and forwards changed the flash from orange to blue and back. The flash effect, again as with fracture-filled diamonds, can be seen only at a very steep angle. Some of the filled fractures contain gas bubbles, trapped and flattened, seen where the filling was incomplete.*

An ingenious imitation of emerald was shown to the GIA by Thomas Chatham of Chatham Created Gems and reported in the summer 1992 issue of *Gems & Gemology*. A specimen reported earlier appeared to have been produced by sawing a light-coloured beryl in half, excavating the two halves and filling them with a viscous green fluid, finally reassembling the piece. When a cutter sawed through the crystal, a green fluid leaked out.

Tom Chatham's specimen, purchased in Bogotá in 1991, was a hollowed-out hexagonal prism which was of a medium dark green colour even without the filler. The colour in fact arose from a coating adhering to a large proportion of the internal excavated surfaces: these appeared colourless when no coating was present. The second component was an apparently water-worn elongated subhedral crystal made from or coated with a green substance. This piece had been inserted in the hollowed-out cavity of the hexagonal prism. This component was probably a plastic judged by its low heft and its softness. The third component was a cap and was also assembled: it consisted of a soft grey metal, perhaps lead, covered by a mixture of ground mineral material in a possible polymer groundmass which melted when the thermal reaction tester was gently applied. This material was reportedly being produced in a Bogotá factory. While most dealers do not come across gem-quality emerald crystals, emerald specialists have to be cautious in their purchasing. *The absence of a chromium absorption spectrum would be suspicious in both of these cases.*

A string of 'aquamarine' beads displayed at the 1992 Tucson Gem & Mineral Show consisted of some pale-blue aquamarines and some colourless beryls, the blue being obtained from coloured thread and dye in the drill holes. When, as reported in the Summer 1992 issue of *Gems & Gemology, the beads were soaked in acetone, most of the dye was removed.*

The ICA Laboratory Alert No. 48 of 5 November 1991 drew traders' attention to the possibility of Pakistan emeralds being fracture filled. *One stone was found to contain dendritic patterns in the filled fractures, the fillings showing a strong yellow fluorescence under LWUV.*

Beryl coated to resemble emerald is probably not manufactured so much today as synthetic emerald is so easily obtained. However, the old specimens do not simply fade away! A 4.39 ct emerald-cut green stone examined by GIA and featured in the spring 1993 issue of *Gems & Gemology* had a green coating covering the entire pavilion. *Through the Chelsea filter the stone appeared red! The properties were those of natural beryl, and inclusions characteristic for beryl could be seen in the substrate. The coating had worn on the facet junctions and melted when touched with the thermal reaction tester. A dye-related absorption band could be seen between 690 and 660 nm with the hand spectroscope.*

In the winter 1992 issue of *Gems & Gemology* the GIA report the uncommon occurrence of stones consisting of synthetic emerald overgrowth on faceted near-colourless natural beryl turning up set in a piece of jewellery: they are almost invariably loose stones when seen. In the piece examined by the GIA the four stones showed the characteristic crazing consisting of fine intersecting fractures at the junction of emerald overgrowth and the beryl core.

Surface-reaching fractures in emerald are now routinely filled with transparent colourless substances. *The filled fractures are sufficiently thin for gas bubbles in the filler to appear flat: under oblique lighting they are highly reflective.* In the winter 1993 issue of *Gems & Gemology* the GIA note an emerald of 8.02 ct in which a large cross-sectional area had been filled. The stone contained three large etch channels with several smaller ones, all of which had been filled and apparently sealed at the surface. *Sealing was found to be inadequate because heat from the microscope lamp caused some of the filler to sweat out of the larger channels. That part of the filler below the hardened surface remained liquid since large spherical gas bubbles could be seen: these were found to move when the stone was rocked on the microscope stage. Orange flash effects were also seen, suggesting that the stone had been treated with a synthetic resin with an attempt to polymerize the surface.*

In the fall 1993 issue of *Gems & Gemology* a piece of glass measuring 22.89 × 17.38 × 12.19 mm, weighing 33.49 ct, was offered as Zambian emerald by a salesman in Zaire. This piece turned out to be green glass with pulverized orange-brown limonite and dark biotite mica flakes stuck to it.

The stability of Opticon as a fracture filling in emerald has been questioned by some treaters who have added hardeners in varying amounts. This has increased the RI from 1.545 to around 1.560.

Though composite stones imitating emerald have been mentioned in the gemmology textbooks for many years, it would be dangerous to assume that such products have been displaced by the synthetic emerald. A note in the spring 1994 issue of *Gems & Gemology* reports that the commonest emerald-like composite is the synthetic spinel triplet with a green cement joining the sections, while similar triplets made from quartz or colourless beryl are frequently encountered. Since the beryl triplet, where the crown and pavilion really are colourless beryl or very pale aquamarine, shows beryl constants, gemmologists have to be careful.

A German firm was reported to be marketing beryl triplets in a very good emerald colour. In this product, care had evidently been taken to use beryl pieces with characteristic inclusions, and some pieces had been carved as cameos. The same firm was offering beryl triplets with a saturated, slightly greenish-blue colour as an imitation of the blue Paraíba tourmaline.

An emerald imitation marketed under the name of Swarogem was announced by D. Swarowski & Co. of Wattens, Austria, in 1994. The stones tested by the GIA and described in the summer 1994 issue of *Gems & Gemology had an RI range of 1.608–1.612 (the manufacturers reported that the RI range was 1.605–1.615). The specimens were isotropic and conformed to the manufacturer's advertised SG of 2.88–2.94. There was a weak green fluorescence under SWUV and a weak to faint yellowish-green to greenish-yellow response to LWUV. With the hand spectroscope an absorption band could be seen centred at about 590 nm with other bands at 483, 472 and 466 nm, these forming a triplet. A strong broad doublet could be seen at about 448 and 442 nm. The stones remained green when observed through the Chelsea colour filter, and were inclusion-free.* The promotional literature gives a dispersion figure of 0.030 and a hardness of about 6.5. Energy-dispersive X-ray fluorescence (EDXRF) analysis showed the material to be a calcium aluminium silicate, but praseodymium was also detected, this accounting for some elements of the absorption spectrum. Swarogem is proved to be a glass with a higher hardness and RI than most ornamental glasses.

In the summer 1994 issue of *Gems & Gemology*, the GIA staff report on an Israeli-produced system (LubriGem) for the fracture filling of emeralds. The equipment comes in two sizes, one size allowing treatment of several 'cups' of about 1000 stones total at a time, the other size consisting of a single cup which holds 50 stones. A later medium-sized development has a capacity of 400 stones. The equipment incorporates a vacuum pump to remove moisture and air from fractures before filling under pressure commences.

Another Israeli system for fracture filling is described in the same report. This one VPO (Vacuum and Pressure Oiling System) is said to be able to treat 'thousands' of stones in a few hours, using a wide variety of natural and synthetic oils and resins.

What appeared at first sight to be an emerald necklace with unusually symmetrical beads turned out to be composed of plastic-covered beryl. This item is reported in *Gems & Gemology* for fall 1995. *Around each of the drill holes was a dark green ring containing flattened gas bubbles.* Closer examination showed that the coating showed red through the colour filter and fluoresced chalky yellow under LWUV. It was shown to be a plastic by its reaction to the hotpoint. The beads were tested and found to be natural beryl.

A short note in *Gems & Gemology* for spring 1995 mentions that hydrothermally grown synthetic emerald grown in Siberia has been offered as 'Maystone, Siberian-Created emerald'.

A report in *Gems & Gemology* for spring 1995 mentions a claim by a Japanese firm that they crush Colombian emerald into a fine powder which is then lasered to 'purify' it. Hydrothermal crystals are then grown using the powder as feed. Such a product would have to be classed as synthetic.

The orange and blue flashes now associated with the fracture filling of gemstones were accompanied by additional colours in an emerald tested by the GIA and reported in the spring 1995 issue of *Gems & Gemology. The stone showed fractures extending along its length: across the width of the stone the fractures showed an orange to pinkish-purple flash. Along the length the same fractures showed a blue and orange flash. Also seen in the stone were some irregular and highly reflective bubbles and some whitish cloudy areas in the filler.*

Some synthetic stones manufactured by the Kyocera Corporation have had different characters lasered on to a pavilion face just below the girdle. The characteristics seen so far are a stylized CV (Crescent Vert, a name given to Kyocera emerald) accompanied by the weight of the stone and its quality grade, reports the summer 1995 issue of *Gems & Gemology*. Some 'recrystallized' corundum marketed by the TrueGem Company of Las Vegas has also been lasered, in this case showing identification on the girdle.

Surface-reaching fractures in emerald are now routinely filled with transparent colourless substances. *The filled fractures are sufficiently thin for gas bubbles in the filler to appear flat: under oblique lighting they are highly reflective.* In the winter 1993 issue of *Gems & Gemology* the GIA note an emerald of 8.02 ct in which a large cross-sectional area had been filled. The stone contained three large etch channels with several smaller ones, all of which had been filled and apparently sealed at the surface. *Sealing was found to be inadequate because heat from the microscope lamp caused some of the filler to sweat out of the larger channels. That part of the filler below the hardened surface remained liquid since large spherical gas bubbles could be seen: these were found to move when the stone was rocked on the microscope stage. Orange flash effects were also seen, suggesting that the stone had been treated with a synthetic resin with an attempt to polymerize the surface.*

Red and purple hydrothermal beryl is reported in the spring 1997 issue of *Gems and Gemology*. The material, made in Russia, contains several chromophores and shows pronounced orange-red and reddish-purple pleochroism.

Chapter 10

The quartz family of gemstones

Compared to the corundum and beryl gemstones there is a lot of quartz about. We know it as clear colourless rock crystal, as the intriguing clear smoky quartz, the mauve or purple amethyst, the orange–yellow–brown citrine or the pale rose quartz. There are other less common colours: a pale blue and a distinctive green. Some stones show the cat's-eye or the star effect and others, while transparent, contain sufficient coloured crystals as inclusions to give colour to the whole stone, as in green aventurine.

All the above grow as single crystals, but there are also the translucent to opaque quartz varieties. These include the many different colours of chalcedony and the striped tiger's-eye: we shall meet them later. The whole quartz family is described in my book *Quartz* (1987, Butterworth-Heinemann).

Quartz has a number of properties that make it of exceptional interest to the technologist: we all know the quartz watch but there are many other electronic devices that make use of the piezoelectricity of quartz. This has led to the growth of quartz crystals on a great scale and since growth has been possible for most of this century it is not surprising that some of the crystals have found their way into the gemstone world.

The artificial growth of quartz crystals is carried out by the hydrothermal method, which we have already met in the growth of some corundum and beryl. Very large pressure vessels have been constructed, and growth takes place under highly efficient conditions. The hydrothermal growth method, as we have seen, leaves few traces behind, and the study of much synthetic quartz is not easy. Most natural transparent varieties of quartz, however, will contain mineral inclusions.

Quartz crystals grown for electronic purposes need to be free from twinning, a crystallographic phenomenon in which two or more individual crystals grow together in particular relationships (Figure 10.1). Twinned crystals cannot be used in electronic devices and all possibility of twinned crystals occurring during growth has to be eliminated. This means that ornamental quartz crystals are particularly inclusion-free and consequently hard to detect. While it is relatively simple to grow colourless rock crystal, the coloured varieties amethyst and citrine could not be grown until the cause of their colour was understood, so that jewellery set with these stones is not likely to need testing – as long as their history is fully known! The first gem use of synthetic colourless quartz could not have occurred much before the 1960s, and of the coloured varieties much before 1970.

Growth of quartz crystals was undertaken by many firms across the world, but one of the first to grow it for gem use was Sawyer Research Products of Ohio, USA. I was able to examine a beautiful citrine made by this firm and report on it in 1973: the stone weighed 49.28 ct and showed the normal quartz refractive index and specific gravity. Inside were

Figure 10.1 Natural quartz viewed between crossed polars and displaying twinning. (After P.G. Read)

tiny groups of crystals which may have been the sodium–iron silicate aegirine. This could have formed by a reaction between iron, sodium and silica in the growth process. This historical note shows that fine-quality coloured material has been available for at least 25 years.

While rock crystal is colourless, the colour of amethyst and citrine may arise not only from trace element impurities as with chromium in ruby and emerald but from the operation of colour centres. These involve the presence or absence of electrons belonging to elements present in the crystal: energies such as those present in visible light either return electrons which have 'strayed' from the sites in which they belong or displace electrons from their designated sites. This is a simplification of course and full descriptions of what really happens can be found in Nassau, *Gems Made by Man* (1980), and other books on crystal growth. The moving about of the electrons absorbs wavelengths from visible light, which can then no longer be white but coloured. One effect of this particular cause of colour is that the process may be reversible, and some of the coloured varieties of quartz have been found to fade under certain conditions. Nassau makes the point that the cause of colour in quartz was in fact discovered in the course of crystal growth research. Some simple chemistry can be called upon to explain the colours.

At normal temperatures crystalline silica (SiO_2) is quartz and unless certain other elements are present it will be in the clear colourless form of rock crystal. The elements causing colour are titanium and iron, both very common in nature but rigorously excluded from the crystal growth process if colourless material is required. Ferric iron may give the citrine colour and ferrous iron a greenish colour which is quite characteristic and unlike the green of emerald or peridot. Titanium is believed to play a part in the coloration of rose quartz and, when present as fine needle-like crystals, gives a star effect (best seen with transmitted light). When rutile is present as extremely small profuse crystals, Rayleigh scattering of light gives a blue colour. Smoky quartz and a particularly characteristic greenish-yellow quartz get their colour from the operation of colour centres, and amethyst, too, is coloured in this way. Some amethyst is heated to give the citrine colour, as we shall see.

When colourless quartz is subjected to radiation by, for example, γ-rays from a cobalt-60 source, it will turn to a deep smoky colour in under 30 minutes. For this to happen aluminium has to be present, but this is virtually always the case whether the crystal is natural or has been grown. In synthetic smoky quartz a brownish-green colour is sometimes present, and the smokiness is frequently present in well-defined bands. The smoky colour can be lightened if necessary, but is stable under normal conditions. Heating to 400°C for minutes only would render smoky quartz colourless, but if the process is carried out at a slower rate much smoky quartz turns a greenish yellow before losing its colour entirely. While some have called this material 'citrine' it contains no iron as natural citrine does. Re-irradiation will make the colour smoky once more. Some greenish-yellow quartz may be obtained by irradiating rock crystal, though some smokiness is usually present. Sometimes a blue colour, stable to light, occurs during the heating of smoky quartz.

When quartz contains iron it can be irradiated to give the amethyst colour. Both the yellow ferric iron and the green ferrous iron will give the amethyst colour on irradiation. As in natural amethyst crystals, the colour settles preferentially in certain areas, and growers need to take account of this when orientating the seed crystal. Some smokiness may persist along with the amethyst colour and has to be removed by heating.

Amethyst is reported to fade on exposure to bright light and fading takes place faster if the specimen is heated at the same time. This is not the case with all amethyst, however. Amethyst may be deliberately heated to lighten the colour and if heating proceeds, the specimen may develop the citrine colour or, when ferrous iron is present, a green colour often known as prasiolite. Both amethyst and citrine will turn a deeper colour when heated to 500–575°C.

The colours and processes described above cannot be detected by the gemmologist, who fortunately does not need to, since quartz is relatively inexpensive and the question of disclosure does not arise. What the gemmologist and the jeweller do need to avoid, though, is the persisting use of the name 'topaz' for citrine. Topaz and quartz are not related chemically or physically. It is surprising how this usage still goes on in some countries.

Rose quartz may reach a deeper colour when irradiated, but this is not a common practice: heating may lighten the colour. Some rose quartz has been reported to fade in light while other authorities dispute this: there may be two distinct types of rose quartz.

A very convincing imitation of both emerald and ruby has been made by heating rock crystal and suddenly quenching it in an appropriate dye. By colour alone the best examples could easily deceive, but the characteristic swirliness of the dye inside the stone, as well as the 'wrong' properties gives them away. Polymer filling has occasionally been found in citrine, amethyst, smoky quartz and even rock crystal.

One of the difficult tests which is sometimes needed is to distinguish natural from synthetic amethyst. Several suggestions have been published over the years, many involving the presence or absence of signs of twinning. One method which seems to have gained approval is to examine the specimen between crossed polars or with a horizontal microscope in whose immersion cell the stone can be placed, and then to ensure that the optic axis is parallel to the path of light through the microscope or through the crossed polars of the polariscope. Brazil twinning can be seen in almost all natural amethyst in the presence of interference colours rather than a succession of broad colour bands shown by the synthetic amethyst. Some amethyst manufactured in Japan has been found to show arrow- or flame-shaped structures: the Gemological Institute of America (GIA) have reported acicular crystals and spicules capped with quartz crystals. Some examples, probably originating in Japan, have been found to show two-phase inclusions resembling feathers as well as distinct growth zoning parallel to one of the rhombohedral faces of the crystal.

A quartz variety called ametrine and known to be natural (it has been mined in Bolivia) consists of both amethyst and citrine-coloured sections in the same crystal. This variety seemed to be well suited to synthesis, or rather to enhancement. The aim is to produce a faceted stone divided into equal amethyst and citrine zones, and there are ways in which this can be achieved artificially by differential heating. Identification is not easy, but natural inclusions should be sought as with all transparent quartz varieties.

The translucent but mostly opaque varieties of quartz include many which are dyed as a matter of course. So many names have been in common use for all these varieties that we shall quote only the most accepted few. The name chalcedony is often used by mineralogists as a useful name to cover most if not all these varieties, which are cryptocrystalline – made up of individual crystals too small for the optical microscope to distinguish. Under chalcedony we find agate, jasper, onyx, bloodstone, carnelian and silica replacements of other minerals and of organic substances and other well-known ornamental materials.

Heating in general will turn some of the paler colours to darker shades and some pale colours to a milky white. The chalcedonies with their porous structure take up dyes very conveniently, and this is especially apparent in the agates whose easily seen bands are attractive and unique even without the assistance given by dyeing, which sometimes imparts a garish appearance. This practice has been known since classical times (Pliny states that 'all gems are made more colourful by being boiled thoroughly in honey, particularly if it is Corsican honey which is unsuitable for any other purpose owing to its acidity' – cited by Nassau, 1994). Just before this passage, Pliny says that clever craftsmen are careful to follow up the veins and elongated markings in such a way as to ensure the readiest sale – this is a clear reference to banded material. Many of the chalcedonies contain iron compounds, and if the specimen is immersed in hydrochloric acid they can be dissolved, with the result that a reddish-brown or yellow staining becomes deposited on the inner surfaces. An old practice of adding an iron nail is said to darken the colour. Honey is used because it contains enough acid to enhance the colour of poor material and to produce deep-yellow, brown or black sections. The especially acidic Corsican honey when used in dyeing would need only heating to produce darker colours. The use of other carbon-bearing substances will be found again when we look at opal.

Treatment with honey followed by sulphuric acid was practised in the German gemstone centre of Idar-Oberstein at least as far back as the beginning of the nineteenth century, and since that time the area has been the recognized agate-dyeing centre – many good-quality specimens are found in the vicinity. Acids and heating are used to give the familiar bright colours to agates and to darken other chalcedonies. Nassau (1994) makes the point that had aniline dyes been used, the colours would eventually have faded, but inorganic dyes give stable colours.

In general the dyeing process involves cleaning the specimens to remove grease and then acid treatment to remove iron traces, unless these are considered desirable for preserving a reddish colour. Specimens are soaked in concentrated nitric acid for a day or two with the boiling point being reached: cooling and washing follow. When a red colour is required, iron nails (0.25 kg) are added to the nitric acid solution: by the end of the process iron oxide forms in the pores of the specimens. Green colours are obtained by the use of chromium oxide and nickel oxide has also been used. To obtain black agate (a process passed to Germany from Italy), sugar and concentrated sulphuric acid are used, the acid removing water from the sugar and leaving black carbon behind.

Blue colours are obtained by using either a solution of 250 g of potassium ferricyanide in warm water followed by immersion in a warm saturated solution of ferrous sulphate with some sulphuric and nitric acids in very small amounts. It is said to take 4–8 days for the blue colour to develop. The other method is the use of potassium ferrocyanide with an iron salt. The colour produced by both processes is the well-known Prussian blue. The number of different recipes is great, and one of the best accounts, from which the above is drawn, via Nassau (1994), is *Das Farben des Achats* (1913) by Dr O. Dreher. This book was published in Idar-Oberstein.

Many of the processes and chemicals used in the dyeing of gemstones are very dangerous and should never be attempted in other than strict laboratory conditions and in the presence of someone well versed in the practices.

Dyeing of unbanded chalcedonies is carried out in order to imitate other opaque ornamental materials. Black onyx, popular in such applications as cuff-links, is usually sugar-dyed chalcedony as true black onyx is rare. 'Swiss lapis' is the unfortunate name given to dyed jasper – the blue is unlike that of lapis lazuli, and the SG is that of quartz (2.65) rather than that of lapis (2.83). This is one example where gemmological tests are necessary for the deception to be identified but in general the dyeing of chalcedony is taken for granted and not made the subject of disputes. Banded red and white material may be passed off as natural chalcedony, whatever name may be used, and some material used in cameos (true cameos consist of shell) will be dyed chalcedony.

Sometimes dyeing is used for special effects, including dendritic (plant-like) patterns, which can also be produced by laser, electrical means or chemically: again, there is no real need to be on the look-out for such things. Generally speaking, the colours of dyed chalcedonies are unlike those of the natural material, which are quieter and more attractive to most connoisseurs. Perhaps one tricky example is worth mentioning twice (also under jadeite): some green dyed chalcedony gives a fairly good imitation of green jadeite. In jadeite an absorption band at 437 nm can be seen: this is not seen in chalcedony, which shows no absorption in this area.

The attractive material tiger's-eye is a replacement by silica of the fibrous asbestos mineral crocidolite. Here the fibres have been altered, with the rest of the material, to silica and give the familiar orange–yellow–brown stripes. These can be lightened in colour by bleaching with chlorine: hydrochloric acid may remove the filling of the tubes formed by the decomposition of the original crocidolite, and other substances may be used to fill the tubes to give a different colour, often the blue 'hawk's-eye'.

A recent development with rock crystal is 'Aqua Aura' – crystals or small crystal groups with a thin coating of gold. This gives the crystals a blue colour mingled with interference colours resembling tarnish. Whether or not the intention is to simulate aquamarine, the characteristic form of quartz crystals is obvious, with horizontal striations (grooving) on the prism faces (the long faces meeting in parallel edges) and the pointed rhombohedra forming the termination. The blue colour derives from the transmission colour of gold. A coating of silver or even platinum gives the interference colours without the blue. Rock crystal heated and quenched in water develops cracks from which interference colours may imitate opal: the effect is interesting but not too convincing and in any case the RI and SG of the rock crystal are higher than that of opal, respectively 1.55 and 2.65 compared to 2.1 and 1.45.

Unusual and certainly uncommon treatments of rock crystal include surface coating producing a deep-yellow or a green colour. Sometimes the coating is applied only to the central pavilion facets: this can be observed from the side of the stone if anything is suspected; usually this kind of colour is not quite the one expected. The star effect seen in quartz is often rather faint and has been enhanced by the use of reflective foils. Generally, however, the treatment of quartz is not difficult to detect.

Reports of interesting and unusual examples from the literature

Items in this section have been chosen to illustrate points made in the chapter and to bring one-off items to your notice.

A supposed dyed black chalcedony featured in the summer 1986 issue of *Gems & Gemology* turned out to be devitrified cobalt-bearing glass. By reflected light the specimen appeared black *but by strong transmitted light from a fibre-optic source and viewed under dark-field conditions the piece was blue with semi-translucent edges. Pronounced dendritic patterns inside the specimen confirmed that it was devitrified glass, with an RI of 1.50. A cobalt absorption spectrum was obtained with the aid of fibre-optic illumination.*

Dyed black chalcedony beads are reported in the Winter 1988 issue of *Gems & Gemology.* A broken strand of black beads submitted to the GIA showed some of the beads with high lustre and others where this had dulled. RI readings on both types gave 1.54, which suggested they were all chalcedony. In strong transmitted light the beads with the lower lustre appeared semi-translucent with a brownish-grey body colour, the more lustrous beads remaining an opaque black. In the duller beads, transmitted light showed parallel agate-type banding and a thin black layer which did not cover all the area. This layer was easily removed with an acetone-soaked cotton swab. The necklace was proved to contain surface-coated beads made to resemble dyed chalcedony.

Much black onyx is manufactured by darkening lighter-coloured chalcedony by a sugar–sulphuric acid treatment, but some examples have been reported in which pale material has been darkened by staining and electrolysis: this also produces dendritic patterns which give the pieces extra sales value. Copper salts are dissolved in water to saturation point, after which the chalcedony, already cut to requirements, is immersed for a period in the water. This leaves the specimen a blue–green colour. An electric current is then passed through it, causing the ionic copper solution to break down. The process of electrolysis creates a slowly spreading copper dendrite. This material does not truly reflect any natural counterpart.

It is always hard to spot the sparse inclusions in any gemstone grown by the hydrothermal method since there is no flux to remain behind and no natural solid inclusions. In a synthetic amethyst it may be possible to see nailhead-like spicules with a quartz crystal cap. Look for anything wedge shaped: a useful note can be found in the winter 1986 issue of *Gems & Gemology.*

In the spring 1989 issue of *Gems & Gemology* a rock crystal with manufactured three-phase inclusions is reported. Seen at the Tucson Gem & Mineral Show, the crystal contained water, a gas bubble and a small faceted (!) red or blue stone. They had been introduced into the crystal by drilling thin tubular columns into the stone from the crystal base. The columns were then partially filled with a liquid, and the faceted gemstone introduced.

Also at the 1989 Tucson Show was a flat translucent blue–green cabochon of chalcedony in which a large dendritic inclusion could be seen. This could have been produced by soaking porous chalcedony in a copper solution, then applying an electric current to precipitate out the copper in dendritic form. A small area on the surface of the specimen shows the point at which the current was applied, since the inclusion reaches the surface there.

One of the first reports of the gold-covered quartz known as Aqua Aura is in the winter 1988 issue of *Gems & Gemology.* Both single crystals and crystal clusters are coated with a layer of gold so thin that a blue to greenish-blue colour is transmitted. The quartz is colourless so the colour seen comes entirely from the coating, which is very hard to remove. *The gemmological properties for quartz remain unchanged, and there is little chance of the crystals being mistaken by gemmologists for aquamarine since quartz characteristics (horizontally striated prism, the long predominating faces) are easy to see.*

With the coming of synthetic amethyst, gemstone dealers have so concentrated in looking for signs of Brazil-law twinning that they sometimes fail to spot the occasional synthetic amethyst-coloured sapphire. These are often slipped into parcels of amethyst.

While not very common, rock crystal coloured by quenching and dyeing after heating turns up as emerald or ruby: some examples show a very convincing colour. The winter 1989 issue of *Gems & Gemology* reports the use of the clearly undesirable name of 'green amethyst' for some of this material. *The lens will show the uneven distribution of the dye.*

An imitation of black onyx is reported in the fall 1989 issue of *Gems & Gemology.* The material was in the form of six black drilled tablets *which gave a RI of 1.53. The SG was found to be 2.51, both figures a little low for onyx. The tablets showed a granular surface in the drilled area, an effect consistent with onyx, but some tables showed a glassy conchoidal fracture on the corners, suggesting glass rather than onyx. The hardness was found to be only 5.5–6.* Using a black chalcedony as a control stone, the material was tested by X-radiography and found to be moderately opaque in these conditions. The opacity suggests that the glass contains an element with a fairly high atomic number.

In 1989 the International Colored Gemstone Association reported that 20 000 ct of synthetic amethyst was arriving every month in New York from Korea. Buyers at that time were warned against purchasing amethyst offered for sale at 10–20 per cent of its normal value. Earlier, consignments of 'Uruguayan' (high-quality) amethyst were found to consist of synthetic amethyst. It is more than likely that dealers will not always test amethyst since the value is not excessive and the testing methods are hard to carry out and interpret.

A material with the trade name 'Rainbow Quartz' is described in the winter 1990 issue of *Gems & Gemology.* In promotional literature the makers state that 'molecules of silver/platinum are allowed to adhere to the natural electric charge surrounding quartz crystals. The extremely thin transparent bond breaks light into a rainbow of colours'. *Strong iridescence can be seen on the surface when specimens are examined in reflected light.* The quartz itself remains colourless.

The material known as 'Aqua Aura' is an attractive blue-coated quartz with a characteristic thin-film iridescence. So far the treatment seems to be confined to single crystals and clusters. A thin film of gold is applied to the external surfaces thus allowing blue to greenish-blue transmission from the gold as well as the iridescence. Faceted topaz treated in the same way is reported in the fall 1990 issue of *Gems & Gemology.*

While plume agate (translucent with dark dendritic patterns) would not be considered an expensive material, it has none the less been imitated by an assembled product, as reported in the summer 1990 issue of *Gems & Gemology.* A 61.39 ct specimen was near-colourless and almost transparent with dark reddish to greenish-brown dendrites. *The top was a translucent convex cap glued to a flat, light-grey semi-transparent base. Between the cap and base was a fairly thick colourless layer containing many gas bubbles. This was easily scratched and was probably an epoxy resin.* The specimen was determined to be a glass and dendritic agate doublet.

Dyed green quartz can be a quite convincing emerald simulant. In the fall 1992 issue of *Gems & Gemology* some material current at that time is reviewed. The colour was enhanced by dye-filled fractures, and some of the stones resembled oiled emeralds in their velvety appearance. *When the stones were placed table-down about 3–5 cm above a white background and viewed through the pavilion it was easy to see the green dye-filled fractures. Normal gemmological testing showed that the material was quartz and an absorption band extending from approximately 690 to 660 nm was consistent with a green dyestuff.* With infra-red spectroscopy a series of sharp absorptions could be seen at approximately 2965, 2930 and 2870 cm^{-1}, these being consistent with those found in the epoxy resin Opticon, the material now most commonly used for the fracture filling of emeralds. The GIA believe that a dye had been mixed with Opticon or an Opticon-like resin before being introduced into the fractures of the quartz.

Glass imitations of amethyst *may occasionally give a chalky blue fluorescence under SWUV where natural amethyst is inert.* While this is not an identification, any stone with an amethystine colour and suspicious inclusions could be quickly tested in this way.

Synthetic quartz with a medium- to dark-blue colour was examined by the GIA and reported in the summer 1993 issue of *Gems & Gemology.* The material was doped by cobalt and *showed a cobalt absorption spectrum. Between crossed polars the 'bull's-eye' effect was seen along the length of the prism, an effect usual for untwinned quartz. With diffused transmitted light wedge-shaped zones of darker colour alternating with very light blue zones could be seen (better even without magnification).* A white non-transparent synthetic quartz seen by the GIA at the 1993 Tucson Gem & Mineral Show as a stone weighing 63.69 ct showed milky white in reflected light and yellowish orange in direct transmitted light. *The RI was 1.55, and the material was inert to both types of UV radiation. A faint columnar growth was detectable under magnification. The SG was measured at 2.37, a figure well away from the almost invariable 2.651 shown by virtually all single-crystal quartz specimens.* The porosity of the material may be responsible for the lower figure. Energy-dispersive X-ray fluorescence (EDXRF) spectrometry showed that traces of chlorine, potassium, calcium and iron were present. A blue non-transparent quartz has also been seen. The reduced transparency is probably due to the material, produced in Russia, being an aggregate of minute quartz crystals with parallel orientation.

Synthetic green quartz has a colour like no other green stone. A specimen reported in the winter 1992 issue of *Gems & Gemology* looked more like tourmaline than anything else. *The stone showed the normal properties for quartz but also a parallel green banding and some angular brown colour zoning perpendicular to the green banding. Another specimen of the same material contained tiny white pinpoint inclusions.* EDXRF showed the presence of potassium, iron and silicon (the GIA's reference sample also contained minor amounts of chromium). The material was classed as synthetic green quartz.

Dyed green quartz is a popular imitation of jade. In *Gems & Gemology* for summer 1995 the GIA report a dyed green quartzite (metamorphic rock consisting largely of quartz grains) fashioned into an oval cabochon measuring 30.25 × 15.98 × 5.50 mm, which closely resembled fine-quality jadeite. *The RI found by the spot method was 1.55 (jadeite would be 1.66) and magnification showed dye concentration between the grains. Dye was also confirmed by the characteristic absorption band centred at about 650 nm: this is seen in most dyed jade imitation materials.* By the use of infra-red spectroscopy the presence of a substance similar to the synthetic resin Opticon was established. Since the absorptions in the quartzite were much weaker than those previously reported for impregnated jadeite, it is probable that much smaller amounts of polymer were used.

A glass-coated quartz was reported in the summer 1994 issue of *Gems & Gemology*. A strand of 51 round beads, about 8–8.5 mm in diameter, the beads having a dark violet–blue colour, gave gemmological properties consistent with quartz and examination of the drill holes showed that the coloured layer was thinner and lighter coloured in their vicinity. Some beads showed dimple-like surface depressions, these mainly in the area of the drill holes: they showed the colourless and rough-ground surface of the underlying bead. When a bead was sawn with its surface at right angles to the drill hole, the surface colour layer was found to be only about 0.07–0.10 mm thick. Undercutting showed that the surface layer was softer than the bead and had a hardness of about 5–6 on Mohs' scale. While performing the hardness test the surface layer was proved to be brittle, producing conchoidal chips. It was concluded that the beads were quartz coated with a glass-like, perhaps enamelled substance. EDXRF analysis of the coating indicated large amounts of silicon and cobalt (which provided the colour) with smaller amounts of lead, sodium, zinc, titanium and iron.

A synthetic quartz of an evenly coloured light-blue colour was seen at a gem show and reported in the fall 1993 issue of *Gems & Gemology*. The crystal when examined from the side was seen to be a piece of colourless synthetic quartz which had been grown on a medium-dark-blue seed crystal wafer. The crystal faces were so orientated that the colour from the seed was reflected. *The crystal showed the cobalt absorption spectrum.*

While large citrines, greenish-yellow and yellow quartz are familiar to everyone, stones showing zoned yellow and green are less common, though amethyst zoned with citrine ('ametrine') is seen from time to time. In the winter 1995 issue of *Gems & Gemology* an 8.47 ct stone is described, the question being 'natural or synthetic?'. While gemmological tests diagnosed quartz with no difficulty, neither the gemmologist nor jeweller will have easy recourse to EDXRF, or to infra-red and UV-visible spectroscopic techniques: such instruments are essential when quartz origin is in question. *The gemmologist will be interested to know that most synthetic citrine often shows colour zoning in planes at right angles to the optic axis (you have found the optic axis in quartz when you see rainbow colours between crossed polars on the polariscope); scattered inclusions now traditionally known as 'breadcrumbs' are also an indication of artificial origin.* EDXRF gives sharp peaks in the infra-red region near 3000cm^{-1}, and peaks are seen at 487, 458, 420, 398 and 345 nm in the visible-UV spectrum: the peak at 487 nm may be due to trivalent cobalt.

A colour-zoned synthetic amethyst reported in the fall 1995 issue of *Gems & Gemology* showed a very uneven colour distribution, resembling speckles or 'leopard spots'. This is unusual for amethyst, even though this stone routinely shows patchy colour distribution. In this case the cause is thought to be growth on a seed plane cut perpendicularly to the *c*-axis. Amethyst coloration prefers some faces of the crystal to others and settles preferentially in the faces of the positive rhombohedron. Under magnification it could be seen that the darker-coloured regions grew as expanding rhombic pyramids from point inside the crystal. *So far, such a speckled appearance, if ever encountered in a faceted stone, should be taken as proof of artificial origin.*

Chapter 11

Opal

Opal with its play of very pure spectrum colours is one of the costliest of gemstones, especially in its black form where the play of colour is seen against a black or dark background. As the colours do not result from selective absorption but from diffraction, they are pure and bright: it is not surprising that many attempts have been made to synthesize or at least to imitate opal.

The composition of opal is silica with a variable amount of water: opal has no crystal structure since its constituent atoms do not make up a regular three-dimensional structure. Solids with no regular atomic structure are called amorphous (i.e. non-crystalline) and can show no directional hardness or pleochroism: nor may they be birefringent. The porous nature of opal makes tests involving liquids inappropriate since damage or unsightly marking may result, so that the gemmologist has to rely on the microscope to show a close-up of the structure of the coloured patches. The spectroscope is not useful in the testing of opal.

The beauty of opal has earned it many imitators, many of which are made from glass. However, since there is a synthetic opal (perhaps more strictly an imitation) we shall examine this first.

The secret of making a material looking like opal is to find out what causes the play of colour and then to reproduce it in a solid which will withstand ornamental wear. Surprisingly the gemstone world had to wait until the 1960s before the process was finally established. Fittingly, the secret was discovered in Australia, the home of the finest opal.

Simply described, precious opal (the kind with the play of colour – most opal is common opal) consists of a three-dimensional array of silica spheres which, together with the voids between the spheres, diffracts white light in the same way as a diffraction grating. The regular stacking of the spheres is critical since if this is not present no diffraction can take place. Also critical is the size of the spheres: the array in which they are contained has to show a spacing of a similar distance to the wavelength of visible light, so that spheres spaced at a distance of approximately 500 nm will give blue–green light. Looking more closely at the array of spheres and remembering that the composition of opal includes both silica and a variable amount of water, we find that the compositions of the spheres and of the voids differ a little from each other. The difference in composition between spheres and voids produces the diffraction from either the spheres or the voids: a large difference in refractive index (RI) would prevent diffraction and result in a whitish opal since light scattering would take place randomly. This type of opal has been called hydrophane since it shows a play of colour only after it has been wetted: after drying it resumes its whitish appearance. We have already seen that an irregular array gives rise to common opal.

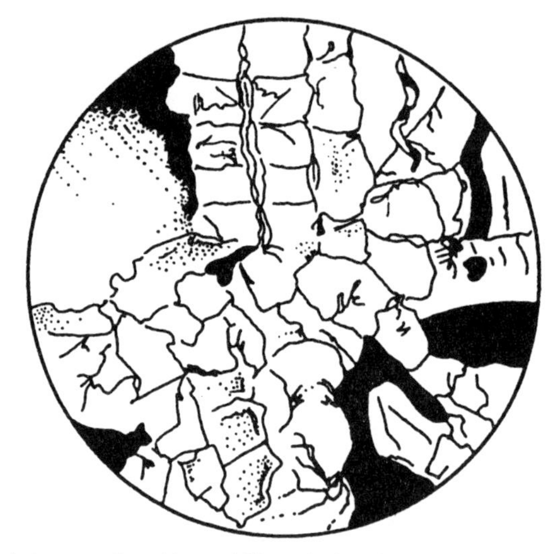

Figure 11.1 Characteristic hexagonal markings within a single colour patch in a synthetic opal

To synthesize opal the regular array of spheres of appropriate size has to be achieved. While quite a number of substances can be made, most are unsuitable on account of their inability to stabilize into a hard material. Silica spheres are the ones to work with and first of all their size has to be uniform. Their subsequent packing into an array with the voids filled up with material which can also be hardened has proved a difficult task for the would-be synthesizer.

The exact nature of the process has not been disclosed by the manufacturers but the aim is to produce an opal with both black and white backgrounds and with as fine a play of colour as possible, in which all the colours of the spectrum can be seen.

One of the first to manufacture opal successfully was Pierre Gilson of emerald fame. Gilson's white opal was first put on the market in 1974 and was followed by the black opal.

Gemmologists are now familiar with the characteristic 'lizard-skin' appearance of the individual colour patches in Gilson and other synthetic opals. When magnified, a subhexagonal patterning can almost always be seen as a background to the colour (Figure 11.1). Gilson opal shows columnar structures arranged at right angles to the base of the cabochon so that when a specimen is examined from the side the colour patches can be seen to extend right down to the base. If synthetic material is used in a doublet or triplet, as it often is, the columns are either too short to see or cease abruptly.

Natural opal may occur either as nodules or, perhaps more commonly, as thin coatings on a sandstone base (matrix). The coating is often both fine in quality so that it cannot be wasted and also very thin, like butter on bread. The problem is resolved by using both opal and matrix as a doublet or, with a domed transparent cover, as a triplet. The domed cover acts as a magnifying lens and may be made of glass, rock crystal or plastic. While such composites must often have been used to deceive, none the less they are an excellent way of using what is often very fine material.

Opal doublets can be detected most easily by their flat upper surface which should arouse suspicion. Then examination from the side should show the sudden change in appearance. Opal should never need to be tested any other way and certainly never in a liquid. Since black opal is particularly valuable, the base of a composite will often be either darkened matrix or some other dark material, perhaps glass. A black onyx base with a synthetic corundum cover perhaps represents the ultimate triplet: plastic covers are easily scratched and neither rock crystal nor glass tops are encountered every day.

The name 'treated opal matrix' is given to a whitish opal with a play of colour, which has been darkened by the presence of carbon as tiny spots (like newspaper pictures) mingled with the colour. It is surprising that no trade name has ever been proposed for this quite attractive material. The means of getting the carbon are ingenious and varied: old sump oil may be used to immerse the stones and then heated in such a way that the carbon from the oil remains in the opal. Alternatively brown wrapping paper may be used to enclose the opal matrix while heating takes place. The specimens are preformed before treatment takes place.

Opal with no play of colour may be impregnated with colourless material such as oils and waxes. Glycerine, which has sometimes been used, tends to ooze from the cracks or to dry out fairly quickly, so that a specimen which feels sticky should arouse suspicion. One process quoted by Nassau (1994) probably involves drying followed by impregnation in a vacuum. Plastics and silane polymers have been used and plastic coatings of treated opal matrix have also been reported.

Less effective opal imitations usually involve some kind of glass. It may be foiled with a reflective substance or by mother-of-pearl: I have recently seen what appears to be a mother-of-pearl doublet (or triplet) whose play of colour could suggest opal to the unwary, though it looks much more like mother-of-pearl. Genuine opal may be coated on the back with a dark substance to improve the appearance of the play of colour. One interesting example of an imitation makes use of a dyed fish-skin (Schnapperskin triplet). Another uses fragments of opal with a play of colour immersed in a liquid (glycerine or even water) in a plastic housing.

Most opal imitations can be detected by magnification or, when coated, by signs of oozing when the thermal reaction tester is brought close. A needle will scratch plastic in a characteristic way which is unlike the reaction shown by opal. Nassau (1994) reports that a Brazilian plastic-impregnated opal has been found to contain tiny crystals of a nickel–iron sulphide.

While opal imitations are most realistic when formed from silica, other materials have been used. One interesting example is made from latex, which can be induced to form the appropriate array of spheres but which has the very low SG of 1.0 with a perceptibly light feeling; another plastic is used for hardening, and this has a different RI. An infra-red spectroscopic study of another opal imitation showed that it was made from a co-polymer of styrene and methyl methacrylate. This gave an RI of 1.465, while yet another had an RI of 1.48 with an SG of 1.17.

After handling a large number of synthetic opals, the sense of a blush of a single colour sweeping across the stone as it is moved does suggest an artificial rather than a natural product. A very early Gilson product showed narrow coloured stripes along the long direction of the flattish oval cabochon: this when showed to some London dealers in the early 1970s provoked some immediate offers! This type of pattern was not repeated so far as I know. The stone also showed surface markings which suggested some form of exudation.

The now well-known Slocum stone, for which the name opal-essence was used at one time, is a glass containing small thin fragments of laminated material such as tinfoil, and manufactured by a controlled precipitation process. It has an SG of 2.4–2.5 with an RI of 1.49–1.50. With the lens, gas bubbles and other signs of glass can easily be seen. While many examples of Slocum stone (named for the inventor, John S. Slocum of Rochester,

Michigan, USA) are too gaudy (though attractive and unusual) to be taken for natural opal, I have seen some specimens which are quieter and quite convincing in an opal imitation role. A paper in the *Journal of Gemmology*, Vol. 19, P. 7, (1985), suggests that some of the colour flakes may be made from remnants of large single sheets produced by a sedimentation process. In two other opal imitations, made this time from plastic, one showed none of the lizard skin effect, while in the other the colour domains were scattered to give a fairly good imitation of natural opal. The main constituent was colourless polystyrene, but in both specimens an absorption region was seen at 590–565 nm, perhaps arising from some aspect of the structure. Specimens had an SG of 1.18 and an RI of 1.485: a strong bluish-white fluorescence could be seen under long-wave ultra-violet radiation (LWUV).

In recent years the Gilson process has been used to produce fire opal with an orange body colour. Some specimens show a play of colour; others appear to be without it. Gilson has also produced a water opal: this name is used for transparent colourless opal in which the play of colour appears to hang in colour patches as if in water. Both natural and artificial water opals are very beautiful.

Summary

The name 'synthetic opal' may be used for specimens which should more accurately be called imitation opals, but to try to enforce this is perhaps rather pedantic. Natural opal has an SG of 2.10 and an RI in the range of 1.44–1.47: the hardness is usually approximately 5.5–6.5. The various types of synthetic opal often approach these figures but rarely seem to exceed them, while some specimens give distinctly lower ones. Some natural opal will fluoresce yellowish under LWUV, but this cannot be taken for granted nor can the test be used to distinguish natural opal from its imitators. Ruling out most standard gemmological tests because of the porous nature of opal, we are, as so often, forced to examine specimens under the microscope: here the lizard skin effect is seen in most synthetic specimens, and imitations made from glass show gas bubbles and swirls. Opalite, a very common and long-established glass, despite the name, is used for the backing in black opal composites: these are formed from thin slices of both natural and synthetic opal and should be checked from the side as well as for the lizard skin effect.

Imitation and synthetic specimens apart, the background appearance of opal can be improved to enhance the play of colour, always darkening it to give the effect of black opal. Carbon from a wide variety of sources can be seen as dots among the patches of colour in the oddly named treated opal matrix.

Reports of interesting and unusual examples from the literature

Items in this section have been chosen to illustrate points made in the chapter and to bring one-off items to your notice.

An opal doublet with milky white opal on a dark blue sodalite (both materials from Brazil) is reported in the summer 1996 issue of *Gems & Gemology*.

An opal which had been both sugar treated and coated is reported in the fall 1990 issue of *Gems & Gemology*. The stone had a uniformly black body colour with a fairly strong and evenly distributed pinfire play of colour, with green predominating. While the RI of the dome was 1.45, the base gave 1.56–1.57, under LWUV the dome fluoresced a very strong yellow–green while the base glowed a very strong chalky bluish white. Short-wave

UV reactions were similar but weaker, and there was no phosphorescence. The surface could be indented easily with the point of a pin, and the SG was measured at 1.91. The stone had the characteristic 'peppery' appearance of Australian sugar-treated opal: this arises from black carbon spots mingling with the play of colour. The whole stone was also found to be coated with a transparent colourless substance, thicker on the base and containing gas bubbles. Tested with the thermal reaction tester the acrid smell of plastic was noticed.

An assembled chatoyant opal was reported in the summer 1990 issue of *Gems & Gemology.* Opals with a chatoyant band have the effect enhanced by gluing on a colourless cabochon-shaped cap acting as a condensing lens. Opal from Idaho lends itself particularly well to this method of manufacture.

While opalized shell is not too difficult to obtain, its attractive play of colour makes it a reasonable object for imitation. One type is described in the fall 1989 issue of *Gems & Gemology.* Three imitations were examined by the Gemological Institute of America one resembled a clam shell, the second looked outwardly like a mussel and the third like a turban snail shell. They aroused suspicion through their resemblance to dark-yellow brecciated boulder opal. All three specimens consisted of numerous chips of white opal, boulder opal and matrix rock held together with a transparent colourless binder. Laboratory tests proved the binder to be a plastic.

An assembled black opal examined by the GIA and reported in the fall 1988 issue of *Gems & Gemology* was found to be a triplet with a wavy separation plane and a fairly flat top: this was a natural material and showed an uneven surface. An ironstone backing was joined to this surface with a cement tinted to duplicate the colour and appearance of the ironstone. The cement filled the uneven contact surface of the opal top thus giving a single-stone appearance. The joining material melted when lightly touched with the thermal reaction tester.

A Gilson synthetic translucent brownish-orange opal is reported in the summer 1985 issue of *Gems & Gemology.* The near-rectangular specimen showed a very fine play of colour and resembled the best Mexican fire opal. But *viewed from the side the specimen showed thin colourless top and bottom layers with no play of colour. The centre section was brownish orange with different colours from the play of colour confined to distinct areas. Looking through the colourless areas the characteristic structure of synthetic opal ('chicken wire') could be seen. The sections varied in hardness and in their response to UV radiation. The colourless surface area was easily indented with a pin and flowed when the needle of the thermal reaction tester was held directly above it.*

Opalite is the name given to an opal triplet in which a mosaic is used as the opal layer with a clear top and a wax-like base. The mosaic consists of flat pieces of natural opal: in some cases the adhesive used to hold them together was found to phosphoresce. Care should be taken with any opal in a closed setting. A report can be found in the *Journal of Gemmology,* Vol. 23(8) (1993).

Encapsulated Mexican opal using material from Jalisco and shown at the 1991 Tucson Gem & Mineral Show consisted of a slice of colourless, white or orange opal contained in an oval single cabochon (i.e. with a flat back) made from acrylic resin. Manufacture appears to involve pouring some of the liquid resin into a dome-shaped mould, followed by the insertion of a slice of opal with its base coated black to give contrast to the play of colour. Finally, a second thinner resin layer is used as a sealant and to form the base. The stones are said to be manufactured in Guadalajara, Mexico.

A plastic imitation of opal sold in Thailand and reported in the summer 1991 issue of *Gems & Gemology* contained *noticeably spherical bubbles and gave an RI near 1.57 with a strong chalky bluish-white fluorescence under LWUV. With the hand spectroscope, fine absorption lines could be seen throughout the spectrum. The thermal reaction tester produced an acrid smell, diagnosing the material as plastic.*

At the 1992 Tucson Gem & Mineral show a blue enhanced opal was on sale, the starting material said to be Brazilian. The consistency was highly porous and the material was chalky white hydrophane opal with a weak play of colour (in hydrophane opals immersion in water is needed to bring out the full play of colour). According to the spring 1992 issue of *Gems & Gemology* the sellers said that the rough material is first soaked in a mixture of potassium ferrocyanide and ferric sulphate: this produces a dark blue colour. On drying the material is placed in a slightly warmed plasticizing liquid of methyl methacrylate with some benzyl peroxide. This closes the pores and clarifies the opals to near-transparency. The rough is removed and cleaned for fashioning into cabochons before the mixture solidifies. *The blue body is dark enough to appear nearly black. Stones feel plastic, are notably light and show the blue colour under intense light or the lens.*

Some synthetic opal produced by the firm of Pierre Gilson has been found to be porous and to show a better play of colour on immersion in water. A 6 mm round cabochon tested by GIA and cited in the fall 1991 issue of *Gems & Gemology* had a white translucent body colour and a moderate and predominantly orange colour in small angular patches. *On immersion in water the transparency began to increase from the periphery and proceeding inwards. Complete change of diaphaneity took about 35 minutes. The distinctive hexagonal patterning within the individual colour patches became clear as water was absorbed. On removal from the water the stone seemed more transparent and less white than before immersion. When the stone was heated from a microscope lamp it returned to its original appearance.*

Opal is often given a darker background since the play of colour then looks particularly attractive. Many different methods of darkening the stones are known. In the summer 1992 issue of *Gems & Gemology*, silver nitrate is cited as a darkening agent. After the opal is cut but before polishing, it is placed in a silver nitrate solution, which is gently heated for several hours. After cleaning, the stone is heated in a solution of film developer, then finally cleaned and polished. Stones are sometimes placed in direct sunlight between the silver nitrate and the developer immersions. *Specimens are reported to show dark rectangular specks against the play of colour, reminiscent of sugar-treated opals.*

Some types of opal found at Andamooka, South Australia, are known as 'concrete opal': they comprise an opal matrix (precious opal with its host rock adhering) which is softer and more porous than most similar material. Sugar treatment has been carried out on this material which has then been coated with a plastic substance, as reported in the *Australian Gemmologist* for February 1991. *The product is said to resemble the best Honduras matrix opal with characteristic matrix patterns and the expected black spots identifying sugar treatment.*

A glass imitation of opal with the trade name 'Gemulet' is reported in the spring 1992 issue of *Gems & Gemology*. It consisted of colourless glass with small fragments of synthetic opal embedded inside. *The RI of the glass was very close to that of the opal (1.47 and 1.45, respectively) so that the opal has very low relief*, giving a natural impression. The play of colour appears to come from the specimen as a whole. *The material was being marketed both as faceted stones and as teardrop shapes for pendants and earrings.*

Faceted opal with a play of colour is not often seen, but the summer 1994 issue of *Gems & Gemology* reports a range of Gilson-manufactured faceted opals seen at the 1994 Tucson Gem & Mineral Show. The stones were near-colourless and semi-transparent, with a slightly milky body colour. The play of colour included the complete spectrum.

When the stones were examined face-up, most showed a streaky pattern in the play of colour. Such an effect has previously been noticed in synthetic opal cabochons cut with their bases at right angles or at a very oblique angle to the direction of sedimentation of their constituent silica spheres. *For this reason the 'chicken-wire' structures which identify synthetic opal could be seen from the side rather than when the stone was examined face-up.*

An opal imitation made from plastic was on sale at the 1993 Tucson Gem & Mineral Show. Various shapes and calibrated sizes up to 20 × 15 mm are reported in the summer 1993 issue of *Gems & Gemology.* All had a white body colour, and the claims of a promotional flyer were consistent with findings by the GIA on examining sample stones. *A specimen weighing 1.73 ct gave an RI of 1.50 and an SG of 1.17. Under LWUV the stone fluoresced a strong bluish white. The play of colour in some stones was seen as a patchy mosaic rather than the pinpoint effect seen in other opal imitations.* The leaflet issued with the stones stated that growth of 2–3 mm thick pieces with a surface area not exceeding 30 mm square took 5–6 months. Thin pieces have been produced for use as watch faces and also for composite gemstones.

Tests on Russian synthetic opal carried out by the GIA are reported in the summer 1993 issue of *Gems & Gemology.* Among the stones tested were some with a black and some with a white body colour. Many of the white specimens showed some degree of crazing though it was claimed that the dry climate of Tucson, where they were exhibited, had caused the effect in material which had previously shown no signs of crazing. *Testing gave an RI of 1.44–1.45 and an SG of 1.75–1.78 (low for both natural and for other types of synthetic opal – the high water content may be the cause). Under LWUV, stones fluoresced a strong bluish white to blue–white while under SWUV they showed a weak to moderate greenish–yellow to blue–white response.* The play of colour was of the natural opal type, and could only have derived from an opal type of structure. An absorption pattern seen in the near infra-red suggested that organic compounds were present, perhaps acting as a cementing agent for the silica spheres. In the black cabochons the RI was found to be 1.35 with an SG of 1.65. The lower values may have been due to the organic compound.

Assembled crystal opal (water opal in Great Britain and elsewhere) has been found to be made by enclosing an irregular fragment of synthetic opal in glass. Another form, reported in the fall 1992 issue of *Gems & Gemology,* consisted of a transparent colourless top fashioned to resemble a faceted stone with a flat base. The base was foil backed with a diffusion laminate of various patterns, the whole giving an imitation of the play of colour seen in true opal. *The transparent top of one specimen when tested was found to be consistent with plastic, and the rounded facet junctions indicated moulding. With the hand spectroscope a series of dark lines could be seen across the spectrum, some at a slight angle from the vertical and changing their position as the specimen was moved. This effect may be due to the pattern of the laminate.*

An impregnated synthetic opal manufactured by the Kyocera Corporation was seen at the Tucson Gem & Mineral Show of 1995 and reported in *Gems & Gemology* for summer of that year. Six polished freeforms ranging from 3.59 to 4.30 ct in weight each showed different body colours: green, blue, red, yellow, orange and a pale milky stone were offered. The colours were obtained by impregnating the stones with several differently coloured polymers. *Gemmologists will find the material showing red through the Chelsea colour filter and possessing an RI of 1.455–1.470 for polished freeforms and 1.461–1.468 for cabochons. The different colours gave a variety of responses to both LWUV and SWUV: these are unlike any response given by natural opal. The absorption spectra, highly direction-dependent, are also unlike anything shown by the natural stone. The SG is 1.88–1.91, which would be low for natural opal, and any light material with an intriguing body colour should be carefully examined.* The GIA class the material as synthetic opal rather than an opal imitation on the basis of its silica content as established by energy-dispersive X-ray fluorescence (EDXRF) analysis. The exact nature of the plastic used for impregnation has not yet been established. An accompanying report describes the heating of samples of a similar material. The plastic was burnt off at 600°C and the specimens found to be approximately 20 per cent plastic by weight. A chalky residue of high absorbency retained its play of colour.

Fire opal, with its characteristic yellow-to-orange body colour which is sometimes accompanied by a play of colour, has been imitated by a glass, as reported in the fall 1995 issue of *Gems & Gemology.* The GIA laboratory was shown two stones, one a 3.55 ct transparent orange oval modified brilliant, the other a 2.21 ct transparent red emerald-cut. They were reported as having come from a dealer in Mexico. Tests showed the RI values to be respectively 1.522 and 1.480 for the two stones, above the usual figures for natural fire opal, with SG values of 2.64 and 2.39, again above the usual opal SG. Between crossed polars a snake-like anomalous double refraction was seen and small gas bubbles were seen in both stones. It would have been easy for a mistake to have been made with the orange stone since the properties were not too far away from those of opal and the colour was not unusual. The stones were identified as glass. *Gemmologists should look very carefully into supposed fire opals since the gas bubbles could be seen if the observer was looking out for them.* EDXRF analysis showed that the elements silicon, potassium, calcium, manganese and iron were contained in natural Mexican fire opal (the colour is currently thought to be due to disseminated iron oxide or hydroxide particles): in the redder glass imitation selenium was also found. Cadmium has been found in orange and red glasses with similar gemmological properties: many red and orange glasses are known as 'selenium glass' from their cadmium sulphide or cadmium sulphoselenide constituents.

While opal may be natural or synthetic or may be natural material impregnated by plastic or some other substance to enhance strength or colour, there is also impregnated synthetic opal as the GIA found and reported in the winter 1995 issue of *Gems & Gemology.* The opal was produced by the Kyocera Corporation of Kyoto, Japan. The GIA examined a partially polished specimen of translucent black rough with a play of colour. Weighing 3.42 ct, the piece was tough and resembled other treated synthetic opals. With the refractometer (not usually used for porous opal), readings of 1.44 from the side and 1.50 from the top were obtained. Hydrostatic weighing gave an SG value of 1.82: this would be too low for untreated synthetic opal and too low for natural opal, for which the SG normally exceeds 2.00. Like other samples of Kyocera opal, the specimen gave a faint orange reaction to SWUV and was inert to LWUV. The characteristic 'lizard-skin' effect common to virtually all synthetic opal was present, and *this can be seen quite easily with the 10× lens.* Magnification showed that the surface with the higher RI had a thin transparent colourless coating.

The low SG prompted the GIA to test the specimen by Fourier transform infra-red spectroscopy (FTIR). The stone showed several absorptions between 6000 and 4000cm^{-1} which are absent from natural opal but seen in other Kyocera opal: they could be due to a polymer used for impregnation. The GIA, which discloses treatments when they are identified, classed the specimen as impregnated synthetic opal.

Chapter 12

Organic materials

While the majority of gem materials are inorganic, amber, pearl, coral, jet, ivory, bone, shell and others are products of living organisms and thus fall into the area of organic chemistry. Here many of the gem testing methods used for inorganic substances are inappropriate since organic materials are invariably soft, usually porous, and easily damaged by rough handling, by chemicals and cosmetics and sometimes through age alone. Immersion liquids and testing on the refractometer are ruled out and the microscope rightly resumes its place on centre stage. Organic materials are collected as well as worn but there are restrictions in many countries on the import and handling of some materials, in particular ivory and tortoiseshell: such restrictions reflect the greater importance of conservation over ornament.

Amber and pearl have always been problems for the gem-testing laboratory, and if it were not for the relatively low price range covering the majority of specimens a good deal of controversy would be expected. It is not really possible for the jeweller to diagnose the various imitations of amber nor to say with certainty whether every bead in a pearl necklace is cultured or natural: there is not time for such investigations to be made. They have to be left to the laboratory, and it is more than possible that many items go to the public untested since laboratory services are not free. While this is, of course, applicable to inorganic materials as well, the greater prices (usually) paid should ensure that most specimens are looked at more than casually.

The range of organic products is greater than is often realized, and the jeweller may be faced with quite unfamiliar specimens as well as the expected pearls, amber or coral. For reasons not yet explained, almost all gemmologists are happier with inorganic materials, and familiarity with organic ones is quite rare: but when interest is aroused it is usually permanent and developing.

Pearl

Pearls are by far the most commonly used organics, and the finest natural pearls ('oriental' pearls) command very high prices in the saleroom. For this reason they are widely imitated and the imitations are quite successful, even though glass beads in a necklace do not hang in quite the same way as pearls. Imitation pearls are usually some form of glass (very cheap ones may be plastic and feel very light): the real problem is the cultured pearl. Like synthetic ruby or emerald, these *are* pearls – but manufactured by man. The aim is to produce the 'orient' of pearl, an effect which to some extent defies description but which involves a soft lustre with very faint rainbow-like colours in the background. The effect is obtained from overlapping plates (like the slates on a roof), and for pearl to be successfully imitated this structure has to be reproduced. This is why the cultured pearl is the only serious rival to natural pearl.

The structure of pearl and the anatomy of the various molluscs which can produce pearls is well described in gemmology textbooks and will not be discussed here. Cultured pearls were first made by inserting an object into the mollusc, which then covered it with *nacre*, the actual 'pearly' substance, which consists of calcium carbonate and the organic material conchiolin. Early inserted objects were quite large and not only in bead form: figures of Buddha have been used, among other shapes. The presence of large objects can easily be detected by X-ray examination and even by simpler tests, as we shall see. Present-day production of cultured pearls includes the non-nucleated pearl in which no object is inserted: in these pearls part of the mantle (living tissue) from one mollusc is inserted into another, which then commences pearl production. Despite traditional theories, the cause of pearl production is still not understood.

One of the easiest clues to a natural pearl is the shape. While it is not true that an irregular shape always means that a pearl is natural, large irregular 'baroque' pearls are not usually cultured. A necklace of pearls, exactly matching in size, on the other hand, might very well suggest cultured pearl since the chance for making up matching beads is much greater. Another clue, this time distinguishing between natural and cultured pearl on the one hand and imitation pearl on the other, is the smooth feel of the glass imitation pearl surface. Both natural and cultured pearls, with the overlapping platelets making up the surface, feel rough. The only instrument needed for this test is a set of teeth (preferably natural), against which the specimen can be lightly drawn. We should note, though, that there have been reports of rough-surfaced imitation pearls in recent years.

Another test involving little apparatus is to examine specimens in the beam of a strong light. Characteristic stripes may be seen in cultured pearls, providing that the pearly overgrowth is sufficiently thin and that there is a solid nucleus inside. The test, sometimes known as candling, cannot distinguish non-nucleated pearls. A pearl necklace could be tested for specific gravity (SG) but, even though the string makes little difference to the result, the liquid may harm natural or cultured pearls which are both porous. Such a test should only be used to confirm a necklace of imitation pearls: the SG will be less than the range of 2.60–2.78 which covers both types of pearl.

Imitation pearls are usually glass, and the pearly appearance is given by several layers of the natural substance guanine or a synthetic equivalent. This may chip off around the drill-hole making identification easy. The 10× lens will show this effect and also the lack of the characteristic ridgy pearl structure on the surface. If a pin is placed against the surface of a coated imitation pearl the point will sink in when pressed down.

Another form of imitation pearl consists of a glass bead with the pearly effect obtained from a filling. Here the characteristic signs of glass should be visible with prominent bubbles and swirls. This type appears to be much less common than the coated imitations. If the drill hole can be seen, a jagged edge may also serve to identify the specimen, and a pencil point placed on the surface will be seen reflected from the internal surface of the glass.

Other small but useful signs of a cultured pearl are a larger and less straight drill-hole. Pearl drillers will often say that a cultured pearl is easier to drill than a natural pearl. With the 10× lens the nucleated cultured pearl can be identified by a change in appearance when examined down the drill-hole.

There is an obvious limit to the tests which the jeweller can use to distinguish natural from cultured pearl and in all cases of doubt the laboratory needs to be consulted. Tests used by the laboratory almost invariably involve the use of X-rays, and details should be found in gemmological textbooks, where photographs and diagrams outline the apparatus, tests and results.

Black pearls, which command high prices, are really a very dark grey or bronze colour rather than true black. A pearl with a deep, even black colour should be suspected: it may have been dyed with silver nitrate. Such a pearl will not give the dull red fluorescence shown by many natural dark pearls when examined under long-wave ultra-violet radiation (LWUV). If a natural pearl has been dyed and drilled, the silver nitrate may penetrate some of the concentric layers which form part of the pearl structure: metallic silver will be deposited and will appear whitish on a negative photograph taken with X-rays. The print will show the metallic lines as black, of course.

We have to remember that while at one time a black cultured pearl would have been dyed, today naturally coloured black pearls are grown in a number of places. Where a visible nucleus is employed, the silver nitrate may hinder the interpretation of radiographs since it may occupy the junction between the nucleus and overgrowth.

Pink pearls (some may incline to orange) come from the giant conch, and could be imitated by porcelain or by natural (or dyed) coral. Under magnification the natural pink pearls show a flame-like structure on the surface, not seen in the imitations; they also give a silvery appearance when viewed at certain angles under a strong light. Their SG at 2.83–2.86 is higher than that of coral at 2.69.

Other colours of pearl can also be imitated but in most cases the deception is clear. On the other hand some blister pearls (these, in nature, grow partly adhering to the lining of the mollusc shell) have been cultured (the name Mabe is frequently used for them). The domed surface is notably regular, but radiographs clearly show specimens to be cultured. Such pearls were first made by cementing mother-of-pearl beads to the nacreous (pearly) lining of the mollusc: on return to the water the animal produced a dome of mother-of-pearl to cover the bead. The pearls were removed from the shell by sawing and made round by cementing a mother-of-pearl backing to the surface. Today the mother-of-pearl layer covering the bead is lined with wax and backed by a smaller bead, the whole being completed by a domed mother-of-pearl base.

Pearls often give a strong bluish-white fluorescence which may be seen under either LWUV or SWUV. Under X-rays, however, cultured pearls often give a greenish-yellow response while natural specimens respond only rarely. Manganese is thought to be responsible, but the effect is not limited to freshwater pearls. The non-nucleated pearls do not behave uniformly under X-rays either, since Lake Biwa specimens give a strong greenish-yellow response to X-rays while salt-water pearls with the same structure do not seem to respond.

Amber

Amber can cause a great number of problems to the gemmologist and jeweller: specimens are plentiful (though the best ones are expensive) and it is easily imitated. Natural amber can be altered in several different ways and the presence of animal or plant matter inside a specimen is no guarantee of natural origin.

Amber is a *fossil* resin while all other hard ornamental resins are *contemporary*. Knowing this does not always help to distinguish one type from the other though the traditional test for contemporary resins (including copal) is their softening when touched by ether: amber remains unaffected. Fluorescence is too variable to provide a reliable test, and the commoner gemmological tests do not always give a clear result. Colour is not helpful: amber burns or chars with a distinctive aromatic odour while plastics give a sharp sensation in the nose. But the contemporary resins also give an aromatic smell when heated by the thermal reaction tester. Amber imitations made from plastics are sectile (they peel like a pencil being sharpened) while amber (and copal) splinter. Both types of resin and some plastics, when rubbed, will pick up fragments of paper: one of the plastics, casein, will not develop the necessary electrical charge for this effect to take place, so specimens of this milk-based plastic cannot be mistaken for amber by this test.

The SG of amber is usually close to 1.08, and its refractive index (RI) around 1.54, though the material should not be brought into contact with any of the testing liquids. Plastics usually have an SG of over 1.33 and most specimens will therefore sink in a brine solution (50 g salt in 250 ml of water) (copal, with an SG of around 1.06, will float with amber, however).

Insects in amber form a whole subject of study, and without a specialist zoological background no useful information on their identity or origin can be gained. Gemmologists should confine themselves to examining the state of the insect or other occupant of the specimen: if the insect is 'in one piece' and appears rather advantageously placed in the specimen, a strong suspicion should be aroused that all is not well. Insects whose members are detached from the body often indicate natural resin, the flow of which has overtaken them unawares. To tell whether or not the specimen is amber or contemporary resin will need application of the ether test. Amber-like specimens containing larger inclusions, such as small reptiles, will certainly not be amber and are much more likely to be plastic. Grotesque would not be too strong an adjective for some of these pieces, which would, however, not pose too great a testing problem.

Pressed amber (ambroid) has been on the market as an amber substitute for more than 100 years. Pieces, made from fragments of amber compressed together, can be recognized by a characteristic flow structure with elongated bubbles, this arising from the softened fragments of amber being forced through a fine mesh. Bubbles in amber are more commonly spherical. Some pressed amber may not show the flow structure nor the bubbles, however: some at least of these pieces have a slightly lower SG (1.06) than untreated amber and the well-known disc-like structures are also characteristic. Pressed amber of this more recent manufacture has been found to give a strong chalky blue fluorescence under LWUV, and under these conditions a granular surface has also been reported. The ether test does not affect pressed amber.

If the gemmologist or jeweller is faced with amber-like specimens and has no ether handy (quite likely!), it is worth noting that some at least of the contemporary resins may show a crazed surface and that they are perceptibly softer than amber. Copal resin may yield to a knife-blade (careful!).

Plastics are probably the commonest amber imitations, with Bakelite, casein and celluloid seen most often. All have higher SG values than amber and give an acrid smell when approached with the thermal reaction tester. Perspex has an SG near 1.18, while some polystyrenes approach amber more closely with SG values near 1.05. However, should an RI test be attempted, one at least of the polystyrenes will give the high figure of 1.58 (amber gives 1.54): however, any pieces suspected of being amber should not be brought into contact with testing liquids. The polystyrene Distrene is soluble in benzene *but this liquid should never be used*, certainly not by amateurs. As with pearls, the string of a necklace makes little difference if a specimen is tested for SG (using distilled water). A knife-blade, carefully used on 'unseen' parts of large specimens, will show how relatively easily amber chips while most plastics peel.

There is no easy or obvious test for amber, and successful identification is achieved through experience combined with careful use of the procedures outline above. Mistakes are often made, and the gemmologist needs to find out how they arise. This is best done

Plate 13 Sapphire with orange-diffused colour shown strongly at the edge

Plate 14 Natural colourless sapphire prior to X-ray irradiation

Plate 15 The same stone as in Plate 14, now yellow after irradiation

Plate 16 Fine lines in this diamond are laser tracks, aimed at dispersing an unsightly inclusion

Plate 17 Rainbow-like markings in this diamond show that it has been infilled

Plate 18 Unexpected 'foreign' colours are a sure sign of infilling. Green patches in this diamond give the game away

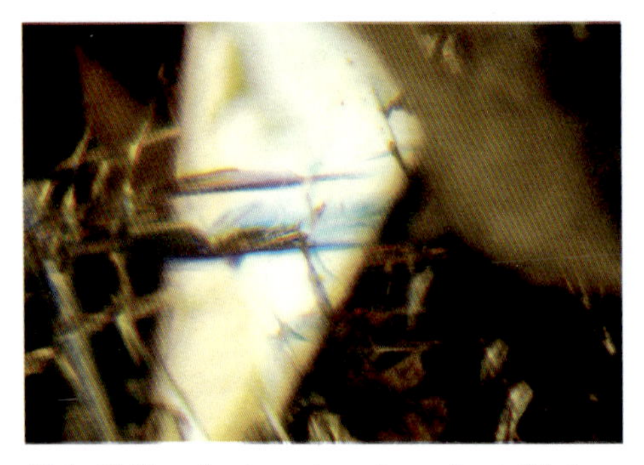

Plate 19 More foreign colours in another infilled diamond

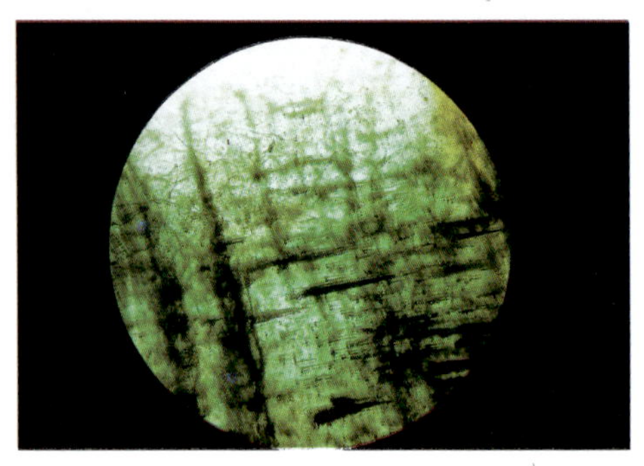

Plate 20 The Nacken flux-grown stone is one of the earliest synthetic emeralds, and is keenly sought by collectors

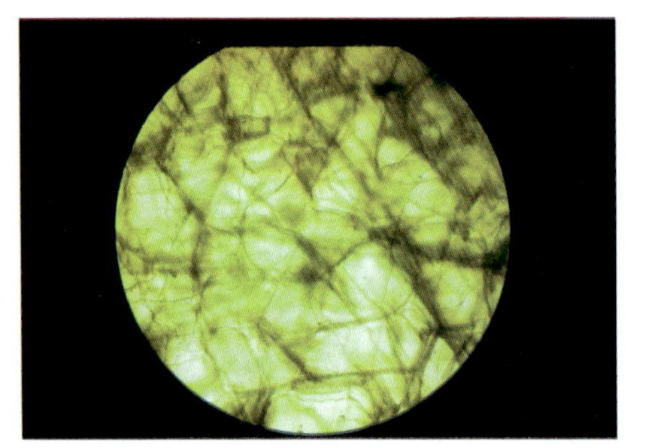

Plate 21 Characteristic veiling in another rare emerald, the Zerfass flux-grown stone

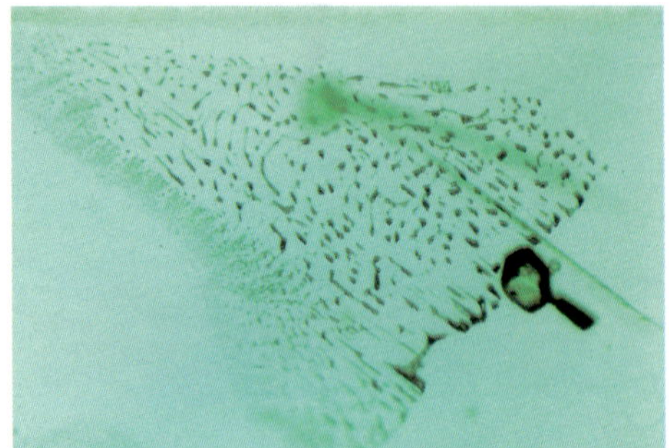

Plate 22 Droplets of flux accompany a crystal, probably phenakite, in a flux-grown emerald

Plate 23 Two-phase inclusions can be seen at the junction of seed and overgrowth in this synthetic emerald

Plate 24 Multiple metallic fragments in a Russian synthetic emerald

Plate 25 Banding in a Russian synthetic emerald

Plate 26 Flame-like structure in a Russian synthetic emerald

Plate 27 Overall scene in a Lennix (French) synthetic emerald. The absence of natural mineral inclusions is always suspicious

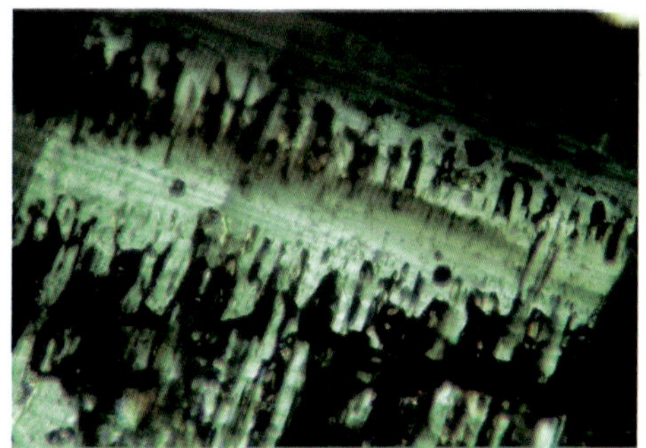

Plate 28 The Lennix synthetic emerald

Plate 29 A Zambian emerald containing tell-tale yellow patches from oiling

Plate 30 Yellowish-brown colours in a resin-infilled emerald

Plate 31 A resin-filled emerald

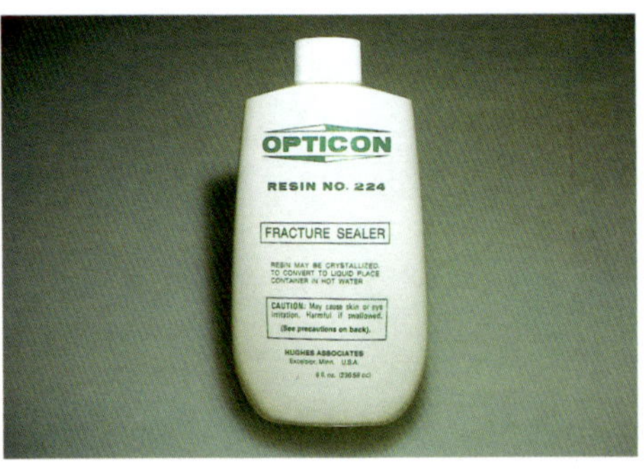

Plate 32 Opticon, the substance of many emerald fillings

Plate 33 Emerald with darker green patches caused by coloured oil introduced through cracks

Plate 34 Lechleitner emerald overgrowth on a natural beryl seed is proved by this characteristic 'crazy-paving' surface

Plate 35 Trails of crystallites in a synthetic green beryl

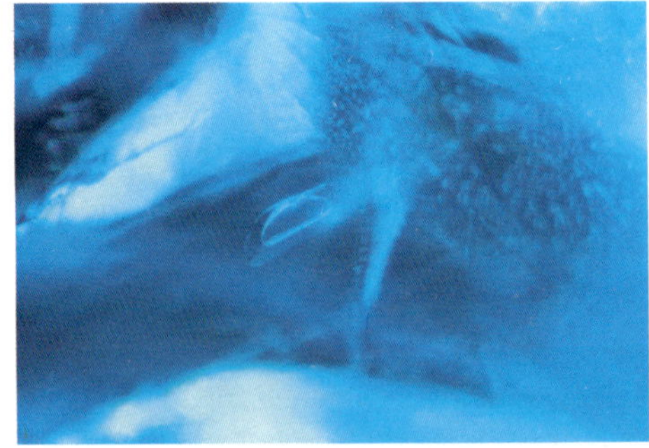

Plate 36 Two-phase and individual liquid inclusions in a synthetic aquamarine

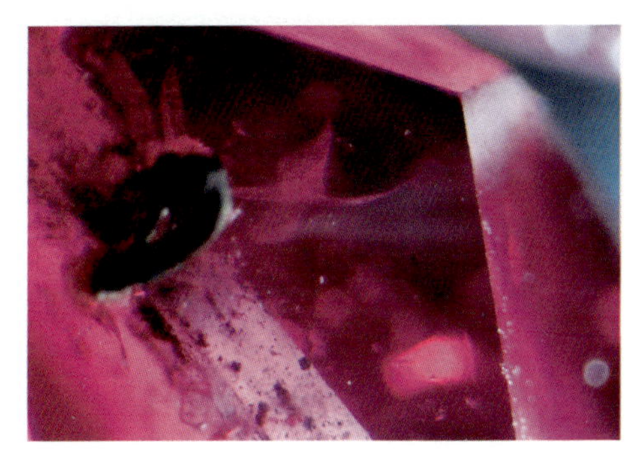

Plate 37 Flux growth is proved by the presence of metallic flakes in this mauve synthetic beryl

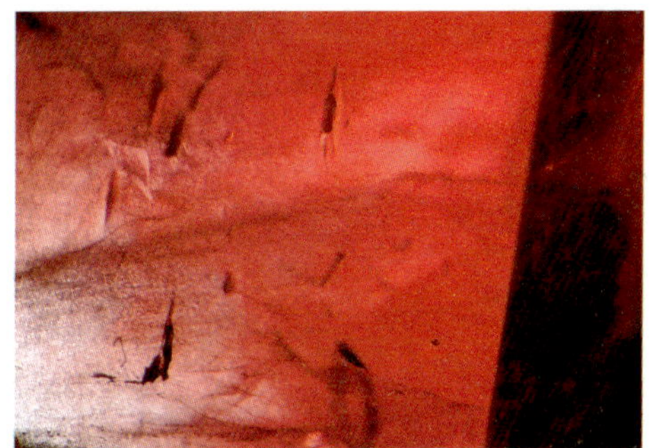

Plate 38 Red beryl is very rare in nature, but the two-phase inclusions in this specimen indicate an artificial origin

Plate 39 Quartz crystals: the blue ones are the coated 'Aqua Aura'

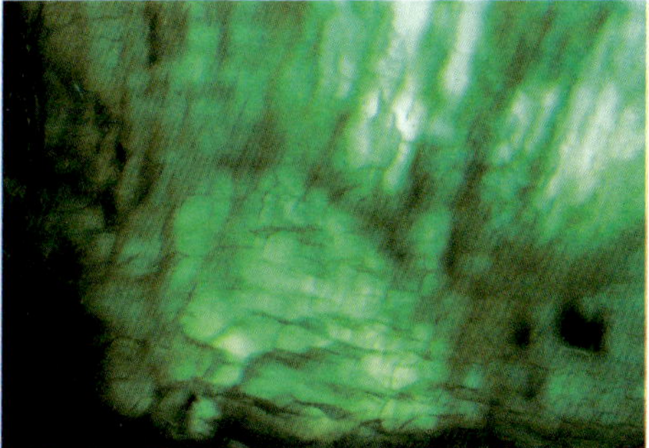

Plate 40 This emerald imitation ('Cirolite') is made by introducing green dye into the cracks caused by thermal shock in a rock crystal. Such specimens are better imitations of emerald than might be expected

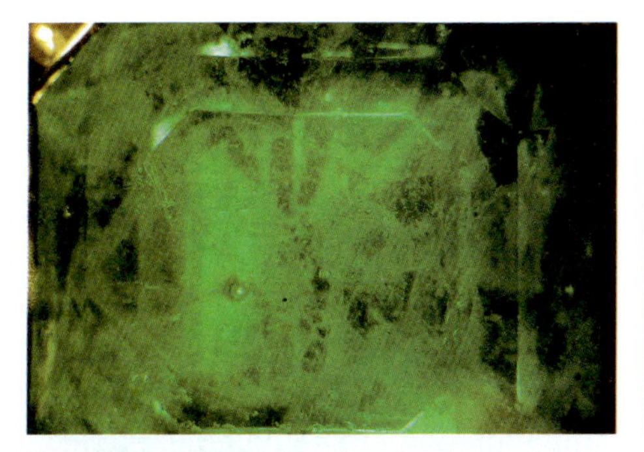

Plate 41 Gas bubbles show that glass is combined with quartz to give this soudé stone imitating emerald

Plate 42 Note that the surface colours stop suddenly when this opal triplet is viewed from the side

Plate 43 No group of natural opals would show such colour uniformity as these synthetic opal beads

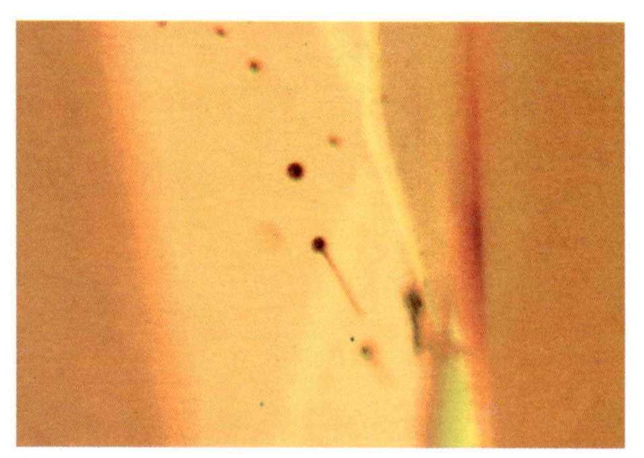

Plate 44 Inclusions in a Seiko synthetic alexandrite – as always, the lack of natural mineral inclusions gives the specimen away

Plate 45 Characteristic structures in a Verneuil spinel

Plate 46 Specimens of colourless untreated and blue irradiated topaz

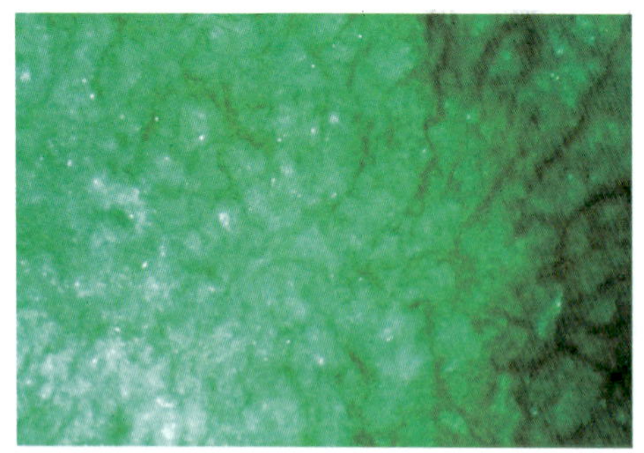

Plate 47 Dyed green jadeite showing characteristic concentration of colour

Plate 48 Plastic-coated jadeite: the plastic can be seen reflecting light

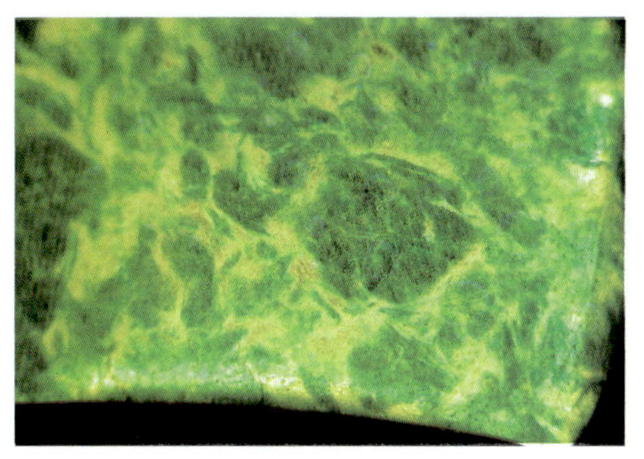

Plate 49 Jadeite cabochon with infilled back

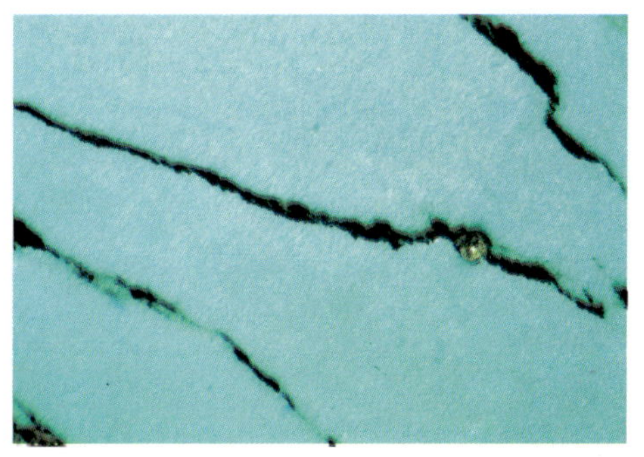

Plate 50 Neolith, a German imitation of turquoise

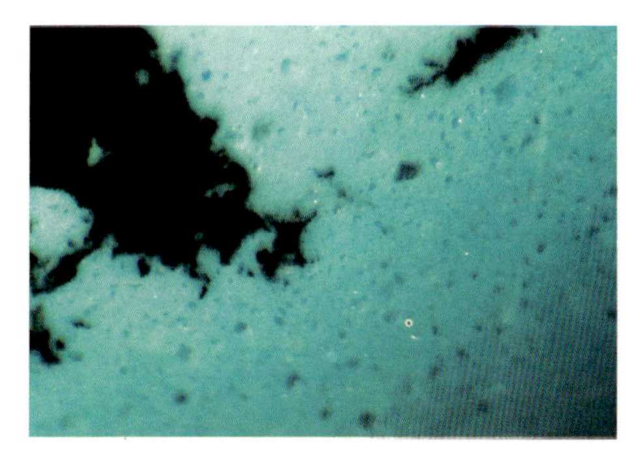

Plate 51 Gilson imitation turquoise with a characteristic surface structure and an imitation dark 'matrix'

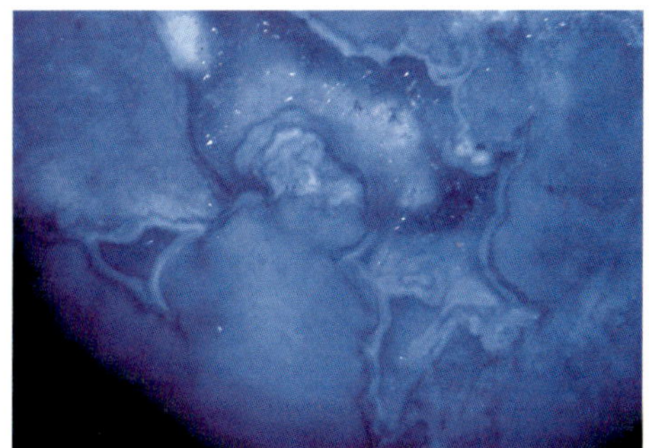

Plate 52 Stained jasper imitating lapis lazuli shows unnatural colour concentrations

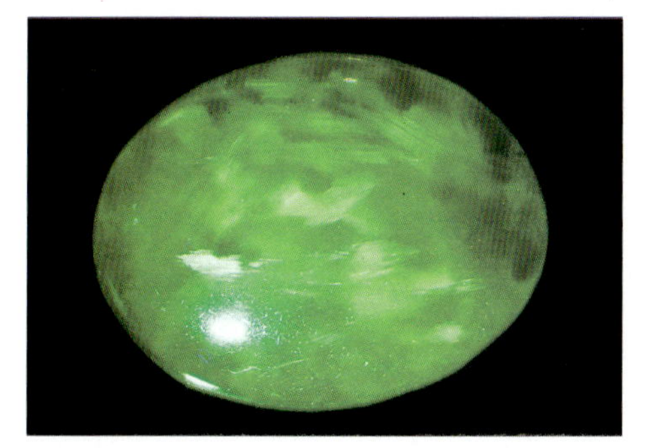

Plate 53 This 'Meta-jade' could easily cause problems until its essentially glassy structure is observed

Plate 54 Another glass imitation of jadeite

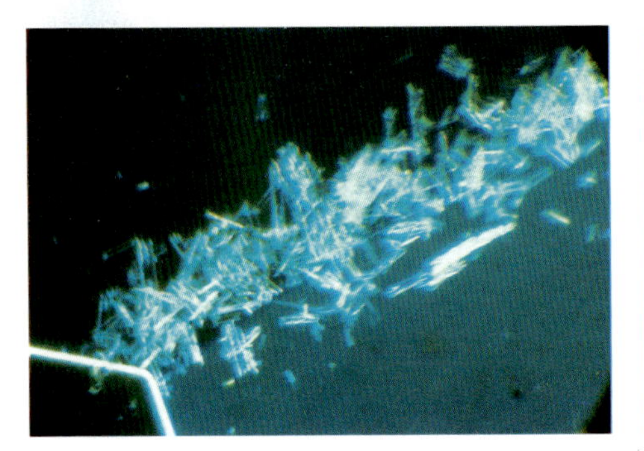

Plate 55 Devitrification can make glass look surprisingly like a natural stone

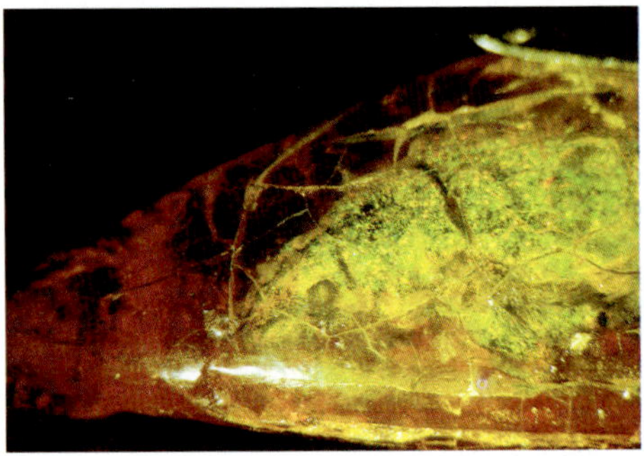

Plate 56 In this amber specimen the beetle inclusion is encased in plastic

Plate 57 Unusual surface colour in a treated amber specimen

Plate 58 Cracking can be seen in this Bakelite specimen

Plate 59 Some of the colours in which cubic zirconia (CZ) is manufactured

Plate 60 Plastic-coated coral

by examining odd amber or amber-like pieces with no commercial value and finding out how they respond to the somewhat harsh tests that may be needed for conclusive results.

Coral

Ornamental coral is distinct from reef-building coral, and is formed of calcite fibres which originate from tree-like branches. The SG is close to that of calcite at 2.68 and the hardness is just below 4 on Mohs' scale. Coral, most popular in its characteristic orange to pink colours, can be imitated by plastics or glass, neither of which will show the coral surface which is made up of ridged structures; the pink conch pearl will not show them either, but is also markedly heavier with an SG of 2.84. Since coral is formed of a carbonate, it will react with a small drop of hydrochloric acid, which will cause the affected area to effervesce (do this under magnification and not in a prominent part of the specimen). Since ornamental plastics do not approach the SG of coral they can easily be distinguished by observing their rate of sinking in a beaker of distilled water. Golden-yellow and black corals are popular, with each having a distinctive structure: golden material shows a spotted or pitted surface while the black material displays concentric rings. None of the natural coral structures can be seen in the coral imitation manufactured by Gilson, probably from crushed calcite: some coral may be dyed, but here the concentration of the dye in cracks and around drill-holes, where present, should show what has happened.

Jet

Jet is fossilized wood, and when heated with the thermal reaction tester will give a coal-like smell. Under magnification it may show a plant- or wood-like appearance. Jet is soft at 3.5 on Mohs' scale, and has an SG close to 1.30. It has been imitated by the artificial rubber vulcanite (which when heated gives a burning rubber smell), and by various plastics. Black glass or natural black onyx will feel colder than jet. Vulcanite has notably rounded edges from moulding.

Ivory

Natural ivory comes from the tusks of the elephant and from the teeth of other large mammals, both land and marine. Fossil ivory comes from the tusks of the woolly mammoth. Whatever the source, ivory has long had imitators, of which bone is the commonest. Elephant ivory can best be distinguished from its imitators by the lines of Retzius: these resemble the fine loops and arcs often seen in engine-turned objects, and are diagnostic. Specimens need to be examined in different directions for the lines to be seen. Other ivories which do not exhibit Retzian lines will show longitudinal striae on the fashioned surface, and in cases of difficulty and with a specimen which will allow it, similar structures can be seen in a peeling examined under magnification, preferably while immersed. The striae or tubular structures are reported to vary with the animal of origin. No ivory imitations show these features.

Ivory has a hardness of 2.5 on Mohs' scale and an SG of 1.71–178 for material from the elephant. This is a higher figure than would be given by plastics. While ivory will usually give a whitish to blue fluorescence under LWUV, not shown by plastics, a similar response is given by bone which has thus to be distinguished from ivory by some other means.

Bone, instead of the lines of Retzius, shows Haversian canals. These are seen as long lines in the material or as black dots when the specimen is examined along them, the blackness arising from surface dirt. While plastics do not show such lines, bone is not usually imitated anyway: it has a higher SG than ivory, near 2.0.

Perhaps the most interesting imitation of ivory is celluloid, which can have an SG as high as 1.90. Celluloid is sectile and smells of camphor (not a familiar smell nowadays) when rubbed.

Other organic materials

Tortoiseshell, trading in which, like ivory, is now correctly forbidden in the UK and in many other countries, may be imitated by plastics. In true tortoiseshell the dark colours can be seen to be made up of reddish dots, whereas in the casein plastics or in safety celluloid the colour occurs in swathes. Tortoiseshell fragments burn with a burning-hair smell (shell and hair are both keratin) while casein when heated gives a burnt-milk smell.

Reports of interesting and unusual examples from the literature

Items in this section have been chosen to illustrate points made in the chapter and to bring one-off items to your notice.

Examples of cultured pearls with repeated drilling carried out to eliminate signs of tissue nucleation are reported in the spring 1996 issue of *Gems & Gemology.* The drillings could be seen in X-radiographs. Predrilled bead nuclei were also found, the drilled nuclei being inserted into freshwater mussels: since these do not open as wide as other molluscs, the drilling enables the pearl farmer to use a special tool for inserting the bead.

An imitation blister pearl cut from the central whorl of the nautilus shell is described in the winter 1986 issue of *Gems & Gemology.* The specimen was believed to be a Mabe pearl or Nautilus pearl, two common names for this kind of shape. *Under magnification, parallel transverse ridges could be seen, an effect not occurring in true blister pearls formed in any of the pearl-producing organisms.* Sections of the nautilus shell are often known as 'Coque de perle'.

An interesting set of blue to grey salt-water cultured pearls is described in the fall 1986 issue of *Gems & Gemology.* The three presumed pearls were 10 mm rounds, two in earrings and the other in a brooch. The grey colour has come to suggest cultured pearls, whether natural colour or irradiated, but in this case X-radiographs showed that each pearl had an opaque centre inside a shell of normal-thickness nacre. Looking down the drill hole showed a white central material with properties like French pearl cement, in that the material appeared to be slightly soluble to the immersion liquid. The problem was: what was the nucleus and why was it no longer there? The drill hole was about four times the diameter of the normal drill hole. The Gemological Institute of America thought it possible that the nucleus had been dissolved after the pearl had formed around it, and that it may have been made from some kind of plastic.

Blue to grey salt-water cultured pearls get their colour from coloured bead nuclei whose colour shows through, or from growth conditions colouring the nacreous layer. They may accumulate an extra-thick deposit of dark conchiolin around the bead nucleus before nacreous deposition begins – again from some abnormality in growth conditions. They may be dyed, in which case traces of the dye will be found around the nucleus and under the nacre – visible through the drill hole. Finally, they may be irradiated, darkening the freshwater shell bead nucleus beneath the colourless nacre.

Cultured pearls in which the nucleus appears dark when viewed down the drill hole may have been irradiated. In a report in the winter 1988 issue of *Gems & Gemology* a well-matched rope of 9 mm grey salt-water cultured pearls was found to have not only dark nuclei but to show enhanced orient as well as improved body colour.

Cultured pearls with wax cores feature in the textbooks, but examples are in fact not too common. In the summer 1988 issue of *Gems & Gemology* a fine necklace of uniformly sized 9 mm cultured pearls is described: all but one fluoresced under X-rays in a dark room. Pushing a pin through the drill hole of this pearl showed that the core was soft and a small portion of it melted under low heat. The pearl surface was nacreous and had the same structure as the other pearls in the necklace. Wax-cored pearls are light and do not hang well.

A pair of earrings examined by the GIA and described in the summer 1988 issue of *Gems & Gemology* contained drop-shaped 'pearls' which turned out to be glass coated with essence of orient. These specimens were accompanied by smaller natural pearls, and it is probable that the glass imitations were replacements. The GIA was uncertain of the date when wax-filled imitation pearls gave way to ones glass-coated with essence of orient.

Dyed black cultured pearls either fluoresce a dull green or are inert to both types of UV radiation, while natural black or grey pearls fluoresce a brownish red (an effect needing darkroom conditions). When examined by X-rays, pearls treated with silver nitrate to darken their colour show a concentration of silver in the area of the conchiolin: silver is opaque to X-rays. In the winter 1990 issue of *Gems & Gemology* the GIA report that a number of grey to black and brown cultured pearls said to come from Tahiti gave a distinct yellow fluorescence under LWUV. From three such pearls examined with X-rays, two showed good contrast between the nucleus and the nacre, while the third had a very thin nacre. A dark fingerprint-like pattern could be seen on the surfaces of two of the beads. Application of colouring material does not need the pearls to be drilled, and the coloured area may be only a thin surface stain. X-ray fluorescence tests showed that tellurium was present in the three specimens.

Cultured pearls containing coloured bead nuclei are described in the fall 1990 issue of *Gems & Gemology*. The beads were reported to be made of powdered oyster shell which had been bonded with a type of cement, then dyed and made into spheres. Beads examined by the GIA were dark greyish green, measuring approximately 8 mm in diameter. The surface indicated an aggregate structure. Near-colourless transparent and opaque white grains were embedded in the spheres. The SG was determined at 2.74 and the RI was just below 1.50. These properties suggest a carbonate, a diagnosis proved by effervescence with a 10 per cent solution of hydrochloric acid. Some green dye was removed by an acetone-soaked cotton swab. X-ray diffraction proved the sphere to be calcite, and the infra-red spectrum showed the presence of a polymer as the bonding agent. *The pearls with these bead nuclei had a notably transparent nacreous layer, and examination of the drill hole shows the thin nacre over the green nucleus. Some of the plastic bonding material melted when the pearls were drilled, and this can also be seen inside the drill hole.* As the nuclei were not made from freshwater pearls no fluorescence could be seen under X-rays. The conchiolin layer could be seen on an X-radiograph, as could the differences between the X-ray transparencies of the nucleus and the nacreous layer.

An assembled and enhanced blister pearl reported in the summer 1992 issue of *Gems & Gemology* had the normal white base of mother-of-pearl but with a top made of a dark purplish-brown nacre with very high lustre and orient. *Under magnification the pearl showed an uneven distribution of colour in the nacreous layer, and with strong lighting and higher magnification irregular dark brown areas were visible. The patches of colour yielded a dark stain to a cotton swab soaked in 2 per cent dilute nitric acid, proving that dye had been used. The pearl fluoresced a dull reddish orange under LWUV (natural black pearls usually give a brownish-red to red fluorescence).*

At the Tucson Gem & Mineral Show held in 1991, GIA staff members noticed a number of white Mabe assembled blister pearls, selling at quite low prices. Most were between 15 and 20 mm in diameter and showed very strong pink overtones, as described in the fall 1991 issue of *Gems & Gemology*. Some showed a spotty, uneven colour distribution which could have been due to enhancement. It was found that a plastic dome had been affixed to the pearl with a very thin nacreous layer upon it. Despite the layer being only 0.25–0.30 mm thick, it was found possible to separate it from the plastic dome, which was found to be coated with a very fine highly reflective material. This layer proved to be a lacquer rather than the usual pearl essence.

While the cause of natural pearl formation is still unknown, those interested in pearls know that the mother-of-pearl bead was not the only nucleus used in early cultured pearl manufacture. The GIA, in the spring 1994 issue of *Gems & Gemology*, report that while a nucleus of white freshwater mother-of-pearl shell was indeed preferred (the bead-cultured 'Akoya' pearls usually have shell nuclei), whole cultured pearls with other nuclei do turn up. Wax and plastic nuclei have been identified, sometimes used with the aim of affecting the overall colour of the pearl. Dyed blue shell cores have been used to give a grey-to-black colour. In one necklace of cultured pearls examined by the GIA a single bead appeared darker than the rest. X-ray examination showed that the nucleus of this pearl was much more transparent to X-rays than the usual shell nucleus. Using the thermal reaction tester a scraping from the nucleus did not easily melt, indicating that a plastic rather than a wax had been used. It was thought that the necklace dated from the 1920s or 1930s, taking into account this particular nucleus, the shape of the pearls in the necklace and the thickness of their nacre.

Treated black Mabe pearls (assembled cultured blister pearls) began to appear on the market during 1992, as reported in the fall issue of *Gems & Gemology* for that year. In one sectioned specimen the nacre top gave a dull reddish-orange fluorescence under LWUV, giving rise to suspicions of treatment. The top was an evenly-coloured dark purplish-brown colour, but a cotton swab soaked in the standard dilute (2 per cent) of nitric acid removed no dye. The pearl was found to have three components: a white mother-of-pearl base, the dark purplish-brown nacre top, averaging about 0.5 mm in thickness, and a dome-shaped core, which in reflected light gave a granular appearance. Under the microscope it could be seen to be made up of translucent white fragments embedded in an whitish mass. A very sharp black demarcation line separated the core material and the nacre top. The core material reacted to a dilute (10 per cent) hydrochloric acid solution. The thermal reaction tester caused the whitish mass to liquefy while emitting the characteristic epoxy resin smell. The nacre top, examined by energy-dispersive X-ray fluorescence (EDXRF), was found to contain calcium with trace amounts of silver and bromine.

Pearls produced by the non-nucleated process in the early years were often shaped like rice grains. Round and near-round shapes can now be achieved, as a report in the summer 1994 issue of *Gems & Gemology* shows. GIA staff were shown a necklace of freshwater tissue-nucleated pearls with beads measuring 6.5–7 mm; incidentally, this size is popular for the bead-nucleated 'Akoya' cultured pearls produced in Japan. When the pearls were

X-radiographed they showed none of the voids whose presence is normally taken as a sign of tissue nucleation.

It is possible that evidence may have been eliminated by the drilling of the hole. Some freshwater tissue-nucleated cultured pearls may therefore not be distinguishable from some freshwater natural pearls if X-radiographs are taken. Reports of the drilling-away of evidence of mantle-tissue nucleation have been published before. It is possible that 'accidental' saltwater tissue-nucleated cultured pearls have been substituted for natural pearls in old pieces of jewellery, probably in the hope that the over-large drill hole would eliminate the voids left by the tissue nuclei.

The name 'Japanese pearls' was used for the earliest products of Japan's work on the culturing of pearls. In the Spring 1994 issue of *Gems & Gemology* the GIA report on assembled cultured blister pearls. In one example the cement plane could be seen, showing that the pearl had a blister top with a cement backing, sometimes with a shell cube or other insert to give stability. In another example a ring was set with what appeared to be coloured freshwater pearls, the colour and shape being characteristic of pearls available at the beginning of the century and popular in the United States. Two of the pearls, on X-ray examination, turned out to be early assembled cultured blisters with the rectangular salt-water shell insert. The pearls had been dyed to imitate American freshwater pearls.

Black pearl has been imitated by an interesting composite, as reported in the fall 1995 issue of *Gems & Gemology*. The item was a strand of dark, silvery, grey-to-black pearls which averaged about 9.44 mm in diameter. The pearl surface had a hazy appearance and a rubbery texture. Under magnification the pearls showed three distinct sections around the drill holes. The inner core was a colourless translucent bead covered by several thin, silvery, grey-to-black layers with another coating of thicker material forming the outer layer. The bead core was found to have a vitreous lustre and conchoidal fracture: X-ray diffraction proved it to have an amorphous structure. The assumption was that the core was leaded glass and the heavy heft of the strand also suggested this. The composition of the thin layers turned out to be a bismuth oxide chloride, the mineral bismoclite which has been shown to be an ingredient of the coating of other imitation pearls, these differing from those imitation pearls coated by a guanine-based substance. The haziness and rubbery effect produced on the pearl surface were due to a coating used to strengthen the bead for normal hard wear. Gemmologists should examine strings of black pearls for signs of different materials being used, looking in the area of the drill hole, and suspect any string feeling unduly heavy.

In the fall 1994 issue of *Gems & Gemology* the GIA describe a black pearl owing its colour to treatment. On the developed X-ray film taken of the pearl (one of a strand of 30 measuring 16.35–11.45 mm) a white ring could be seen surrounding the nucleus, thus proving treatment by a metallic compound which was opaque to X-rays. EDXRF analysis confirmed the presence of a silver compound often used to stain pearls. *Under simple magnification the treatment could be seen to have damaged the nacreous layers, some portions showing iridescence as one effect of this damage. When a pearl is treated with silver salts the entire surface absorbs some of the dye. Dimpling, which is characteristic of many pearl surfaces, can provide a helpful clue since even the dimples will become coloured by the treatment. The presence of a white area in the centre of one dimple revealed a place on the pearl surface where the dye did not reach.*

The continual problem of amber is illustrated by a report in the January 1996 issue of the *Journal of Gemmology*. Here an amber-like box was found to be made from amber by the use of standard gemmological tests: however, some of the ornamental detail was found to be pressed amber by using polarized light. *Between crossed polars and under magnification a pattern of interference colours could be seen with swirly smoke-like inclusions made up of tiny black to brown impurities seen in some places. A reddish-brown dye had been mixed with the adhesive used to attach the pressed amber pieces to the box, and the interference pattern observed between crossed polars could be seen only when viewed through gas bubbles in the adhesive. These dye-free areas showed the original yellow–brown colour of the pressed amber.*

Worry or prayer beads made with 99, 66 or 33 beads with two spacers and a longer cylindrical terminating bead are common in the Middle East and also feature in this report. The set described was found to consist of amber beads, made from amber pieces embedded in a plastic frame: *each bead showed a central line where the two adjacent pieces of plastic adjoined.*

A letter in the January 1995 issue of the *Journal of Gemmology* mentions the darkening of some amber specimens with age. Some pieces are known to show a darker surface and a lighter interior: however, there are some specimens with an apparent colour enhancement of the surface.

Amber is always one of the most difficult gem materials to test and new simulants appear constantly. A 'lac' bead was reported to have been produced from a natural resin at a village (Lino) in the region of New Delhi, India, in the winter 1993 issue of *Gems & Gemology. The beads were opaque with a swirly texture and a yellow to orange-brown colour. Marks resembling the crazing seen on the surface of some fossilized resins could be seen and testing gave an RI of 1.51 and an SG of 1.67.* A section showed a medium dark-brown interior with streaks of a yellow material similar to the colour of the outside of the bead. Under magnification it could be seen that the exterior was really a bright yellow opaque layer with a brownish-orange transparent coating.

Intermixing of the two layers was responsible for the swirly appearance. *Dark brown specks of colour and shallow hemispherical cavities were also noticed, the latter probably arising from gas bubbles breaking the outer layer.*

A letter in the summer 1992 issue of *Gems & Gemology* discusses whether alcohol is harmful to amber. This has been a vexed issue for years, and the writer gives an example from his own experience. When a jewellery display was being set up, a silver ring with amber beads was polished, then washed with soap to remove the polish. A drying rinse in denatured alcohol followed. This caused a hazing to break out on subsurface areas: removal took some hours of scraping, filling and sanding. As far back as 1923 the literature cited alcohol as a threat to amber. Fraquet (*Amber*, 1987 Butterworth-Heinemann) cites scents and hair sprays, which also dull the amber surface.

An imitation of amber is described in the winter 1989 issue of *Gems & Gemology.* The specimen was large and translucent to opaque orange, white, yellow and brown in appearance. One side showed a thin white coating over a brown area in which broken patches showed a third, orange–brown layer. On the other side was more of the white coating with apparently stamped impressions. *The specimen contained numerous gas bubbles and on the same side another break showed a dark-yellow interior. The piece was reported to resemble slag from a plastics factory: neither an SG nor an RI could be obtained. The thermal reaction tester produced the acrid smell associated with plastics,* and this identification was confirmed when an infra-red spectrum was obtained, giving a curve closely matching the standard for polyvinyl chloride.

A simulant of amber consisting of a natural resin in plastic is reported in the summer 1995 issue of *Gems & Gemology.* The material was apparently produced in the former Czechoslovakia from Baltic amber. GIA observers compared this material to other examples of pressed amber, and found that there was *a visual similarity with veil-like grain boundaries visible to the eye.* Recently they have noted the sales terms pressed, reconstructed, reconstituted and synthetic amber applied to another material with some evidence that it was produced in Gdansk, Poland, quite close to some of the classic areas for Baltic amber. Publicity material was said to state that the starting material was 'small pieces of amber taken from the deep ground or from washing up on the shore in the Baltic region ... After being ground they are set in fresh tree sap. After drying they are refinished, polished and hand made into jewellery and other artefacts'. The amber has been sold in many different forms and differs in appearance from true pressed amber by showing very clearly defined irregular transparent to semi-transparent yellow–brown fragments in a lighter-toned transparent yellow groundmass.

Examined by the spot method the RI was 1.56 and the SG 1.24. Between crossed polars, strong anomalous birefringence could be seen accompanied by strain colours. Under LWUV the body of the cabochon tested showed a moderate greenish-yellow fluorescence, and the included fragments fluoresced a moderate bluish-white. The body fluoresced a faint yellowish orange under SWUV while the included fragments did not respond. An acrid smell was produced by the thermal reaction tester but an included fragment reached at the surface gave an aromatic smell. Fourier-transform infra-red spectroscopy (FTIR) gave peaks consistent with natural amber when an amber-like fragment was tested. The matrix gave features consistent with an unsaturated polyester resin. The GIA concluded that the cabochon was a plastic with fragments of a natural resin embedded within it, and that the natural resin was probably amber.

The shortage of natural amber has led to the production of a simulant known by the German name of 'polybern' in factories at the traditional amber centres of Gdansk and Krolewiec. This material is small amber chips embedded in synthetic resin.

Ammolite, the iridescent fossilized ammonite found in Alberta, Canada, has been treated with plastic to reduce the damage inflicted by frost-shattering in some specimens. Sales of treated material began in 1989.

A slab of imitation coral reported in the summer 1990 issue of *Gems & Gemology* was found to be barium sulphate. The slab, weighing 21.26 ct, was an orange–red colour and transmitted a moderate amount of light while appearing opaque by reflected light. The polished side showed a fairly even orange–red colour with a waxy lustre. *When the slab was examined with oblique lighting and magnified, an irregular whitish-pink veining could be seen, as well as black metallic inclusions. The RI was approximately 1.58 and the SG approximately 2.33, with a hardness of 2.5–3.* X-ray powder diffraction analysis determined the material to be barium sulphate.

In the spring 1984 issue of *Gems & Gemology* the GIA report a supposed blue coral. The 17 mm round bead turned out to have a blue core with a near-colourless transparent coating containing gas bubbles. The coating could easily be indented with the point of a pin and gave off an acrid 'plastic' smell with the thermal reaction tester.

Chapter 13

Other species

Alexandrite

Alexandrite is a variety of the mineral chrysoberyl and has the property of appearing red in incandescent light (light bulbs or candle-light) and green in daylight or fluorescent (strip) light. While the colour change in natural alexandrite is not always strong, the phenomenon has always attracted collectors and some jewellery manufacturers, and for the finest specimens very high prices can be secured.

It is not surprising that with its near-unique properties and simple chemical composition (chrysoberyl is beryllium aluminium oxide, $BeAl_2O_4$), alexandrite should be considered as a candidate for synthesis, and this has been carried out for some years using the flux-melt method that we have already met in corundum and emerald.

As the flux-grown material is a synthetic product it has the same composition and properties as natural alexandrite with chromium added to give the colour. Thus the specific gravity would be 3.74 and the refractive index 1.74–1.75 with a birefringence of 0.008–0.010. Alexandrite like other chrysoberyl varieties is hard, more than 8 on Mohs' scale.

In 1973, production of a synthetic alexandrite was reported by Creative Crystals Inc., of San Ramon, California, USA. The material is reported to have been grown by both the crystal pulling and the flux-melt methods but stones entering the commercial market have all shown flux inclusions (Figure 13.1). Alexandrite has also been produced by the Japanese firms Seiko and Kyocera. The stones grown in the USA contain dust-like inclusions and flux veiling resembling smoke: another pulled stone gave a strong red fluorescence under long-wave and short-wave ultra-violet radiations and X-rays. Randomly oriented needles and lath-shaped crystals have been reported.

The laser effect can be achieved with chromium-doped pure Czochralski-grown crystals, and it is surprising that more of these have not entered the market: while some specimens are too pale for gemstone use, others show a good change of colour.

In a report in the fall 1987 issue of *Gems & Gemology* the Inamori cat's-eye alexandrite is described. The material was first marketed in Japan during 1986, and the Gemological Institute of America examined 13 cabochons ranging in weight from 1.04 to 3.31 ct. The stones had a distinct colour change with a broad eye of moderate intensity. Under a fluorescent light source the stones showed a dark greyish green with a slightly purple overtone: the eye was a slightly greenish-bluish-white colour, and overall the stones had a dull, oily appearance. To the eye the stones appeared to contain no inclusions: under a

Figure 13.1 Two-phase inclusions and zoning in an alexandrite grown by the flux-melt method

single strong incandescent light and looking in the long direction, asterism could be seen with two rays appreciably weaker than the eye. This effect has not been reported from natural alexandrite. Gemmological properties were consistent with the natural material but when the stones were placed over a strong light source a strong greenish-white transmission luminescence could be seen – it could also be seen in sunlight and in other forms of artificial light. This luminescence is the cause of the oily appearance of the stones. An unusual weak, opaque chalky yellow luminescence could be detected under SWUV with the effect confined to an area near the surface: a reddish-orange luminescence underlay this. Such an effect has not been reported from natural alexandrite: under LWUV both natural and synthetic stones behave in the same way.

With the microscope parallel striations could be seen along the length of the cabochon. Closing down the iris diaphragm on the microscope showed that the striations were undulating rather than straight growth features: these are not seen in natural alexandrite. In addition, whitish particles oriented in parallel planes could be seen: these planes were associated with the striations and are the cause of the chatoyancy.

Natural alexandrites will contain a variety of mineral inclusions, two- and three-phase liquid inclusions and very visible straight growth features. None of these can be found in the Inamori alexandrite.

For many customers 'alexandrite' is a stone which changes colour between purple and slate blue, often quite large and certainly not expensive. This is a synthetic Verneuil flame-fusion corundum doped with vanadium to give an unusual change of colour. Not only can this material not deceive anyone familiar with chrysoberyl alexandrite, the price asked for the stone is so low that almost anyone should be suspicious. Gemmologists need simply to examine the stone with a hand spectroscope, when a very sharp and prominent absorption line will be seen at 475 nm in the blue portion of the spectrum. Other absorption features will also be present but they are not diagnostic as the 475 nm line is.

While this material is common it looks nothing like true alexandrite. A synthetic spinel imitation is far more convincing, giving a colour change from red to green quite like alexandrite. This is a very rare imitation: despite the confident air of the old-time textbooks I have seen only two examples in nearly 30 years of examining synthetic and imitation gemstones. The gemmologist should be extra careful with any colour-change stone (they are far from uncommon in the mineral world) and check the RI of any specimen appearing to be natural alexandrite. The spinel imitation will give 1.728 compared to the 1.74–1.75 of alexandrite, and will be singly refractive: not only this, but when the specimen is examined between crossed polars the spinel will show a highly characteristic set of dark stripes or bands during a complete rotation, an effect known as 'tabby extinction'; alexandrite would show four times light and four times dark during a complete rotation.

While glass can imitate alexandrite in some circumstances, there is not really a close resemblance. In any case the characteristic features of glass cannot go undetected. A glass imitation was recorded by the GIA in 1973: the specimen changed from an amethyst colour in incandescent light to a steely blue in daylight. This type of glass has been in use for a long time and contains minute crystallites. A glass composite which was made from two pieces of red and green glass cemented together was reported by the GIA in 1965: the two portions were separated by a plane in which minute bubbles could be seen. This specimen showed a different colour in different directions rather than with a change in the nature of the incident light.

We have already noted the synthetic alexandrite made in Japan by the Kyocera Company, who, used crystal pulling to grow 'Crescent Vert Alexandrite' and whose stones were marketed as 'Inamori Created Alexandrite' in the United States. All the synthetic products show either traces of flux when the flux-melt method has been used or have what appear to be featureless interiors if Czochralski pulling is used. The Seiko company, also in Japan, has made alexandrite by the floating-zone crystal growth method; here a crucible is not used. Alexandrite has been made in Russia (very appropriately since the gemstone was first discovered in the Urals). Both pulling and flux growth methods have been used.

The name alexandrite often seems to clear the mind of carefully acquired gemmological knowledge. All the simulants and the synthetics too should respond to tests well within the gemmologist's experience.

Alexandrite has not been colour enhanced as yet, and it is hard to see why it ever should be.

Spinel

To the gemmologist who remembers student days, synthetic spinel is something of a classic. Though a 'genuine synthetic' it does not imitate natural members of its own species but other gemstones. To add to a potential nomenclature problem, the natural and synthetic spinels are not exactly the same since there are compositional differences: these are not great enough, however, to involve questions of species status.

Spinel used as a gemstone is magnesium aluminium oxide ($MgAl_2O_4$) (in nature the spinel mineral group has several members with different chemical compositions). It is easily grown by the Verneuil flame-fusion method, and doping with different elements allows a wide range of colours to be produced. Natural spinel is hard, over 8 on Mohs' scale, has an SG of 3.60 and an RI of 1.718. As a member of the cubic crystal system it possesses no birefringence, and between crossed polars gives a very characteristic and easily recognizable dark-striped effect known as 'tabby extinction' (Figure 13.2): most other cubic minerals normally remain dark in these circumstances during a complete rotation of the specimen.

Figure 13.2 Anomalous double refraction of a spinel grown by the flame-fusion method, as seen between crossed polars

As grown, Verneuil spinels take the boule form. Should anyone ever be asked to distinguish one boule from another (it could happen at a gem and mineral show), spinel boules have a roughly square cross-section and noticeably rough surface, whereas boules of corundum have a roughly hexagonal cross-section and feel less rough. Inside the stones cut from the spinel boule, the curved lines seen in Verneuil corundum are not usually present, and bubbles are seen less often (the growth rate of the spinel boule is slower). Those bubbles that do occur, however, have quite distinctive shapes, some resembling hourglasses or furled umbrellas (Figure 13.3). They can occur in a parallel arrangement, and when viewed in some directions may appear to take up a hexagonal pattern. Flat cavities containing a bubble of liquid or gas may be joined to neighbouring cavities by a tube.

We have already looked at the properties and composition of natural spinel. When crystal growth of spinel was first attempted using the normal composition, some difficulties were encountered, and it was found that by adding extra alumina, growth went better. However, the extra alumina (about 2.5 times) affected the SG and RI, raising them to 3.64 and 1.728, respectively. For the gemmologist and the jeweller with a refractometer this meant that synthetic spinel could be quite easily tested – as long as you thought that your specimen of aquamarine, zircon or peridot was 'not quite right' (these species are frequently imitated by synthetic spinel).

As with synthetic Verneuil corundum, spinel is cheap to grow and provides a virtually endless supply of stones in matching colours – this is why suites of jewellery should be carefully tested if the colours of the stones match closely. Among the colours, red is hard to grow, and stones are quite rare: when seen they show a shuttering or Venetian blind effect which is quite unlike anything seen in any other red gemstone. Interestingly, the red spinels have the normal (stoichiometric) spinel composition rather than the excess alumina. Since the red stones are coloured by the addition of chromium they give a distinctive absorption spectrum in which there are prominent emission (coloured rather than dark absorption) lines in the red: while in some specimens the emission lines coalesce into a single line, in others I have seen a group of up to five lines; textbooks have said that this effect is seen only in natural red spinel.

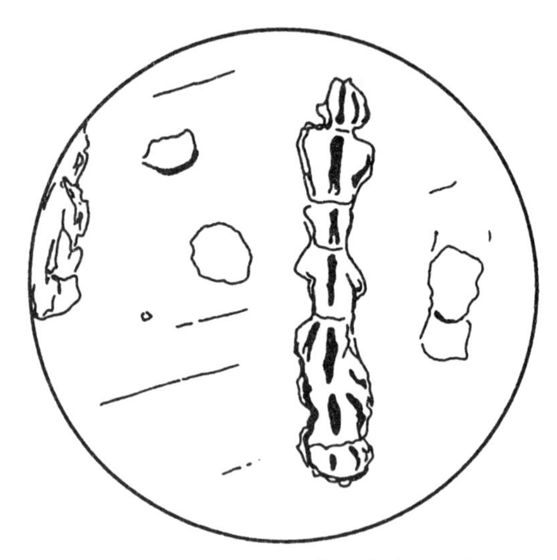

Figure 13.3 Profiled gas bubble in a spinel grown by the flame-fusion method

Manufacturers offer many colours of synthetic spinel: up to 28 from a single source have been reported. RI testing is probably easiest despite the unpleasant nature of the contact liquid, though if you are experienced with the microscope the unusually uncluttered stones should give themselves away. Blue to dark blue synthetic spinel, made by adding cobalt to the feed powder, will show orange to red through the Chelsea colour filter, depending on the depth of colour. If you forget the filter, a cobalt absorption spectrum with broad bands in the orange, yellow and green regions (with the central band the widest – in cobalt glass this band is the narrowest) will be a diagnostic test. A rather vividly coloured yellow spinel fluoresces a very bright lime green under LWUV, but the stone resembles no natural product.

Colourless spinel with no birefringence, quite hard and bright, is an effective simulant of diamond, especially in small sizes. It gives a sky-blue fluorescence under SWUV and until the coming of synthetic gem diamond was the only colourless and reasonably available substitute responding to UV in this way. Particular care has to be taken with large pieces of jewellery set with many small diamonds: up to now those stones which are not natural diamonds will most often have been glass or synthetic spinel. The provenance of a piece has to be established if you are to be sure of not getting a synthetic gem diamond or two! As long ago as 1935 one of the periodical scares gripped Hatton Garden, London, when synthetic colourless spinel was taken to be a synthetic diamond; the answer to this kind of thing is a gemmologically educated trade, but this is rare in all countries. If diamond and synthetic spinel are immersed in a liquid, perhaps di-iodomethane, which has an RI of 1.745, the facet edges of the diamond will be easily seen while those of the synthetic spinel will be almost invisible.

We should also remember that colourless synthetic spinel may form part of a composite since it is hard enough to stand any wear likely to be inflicted on it. A good example of this role played by synthetic colourless spinel is one type of soudé emerald, in which it forms both the crown and pavilion, with an emerald-dyed cement between them.

In the 1960s and 1970s when a great deal of experimental crystal growth was undertaken by universities and research bodies, synthetic flux-grown spinel was one of the materials produced. With a range of dopants, a number of unusual colours were obtained, and crystal groups showing the octahedral crystal form taken by natural spinel were made. Some of the groups might have entered the markets frequented by specialized crystal collectors, but the individual crystals would be too small to facet. While growth of this particular type is not so common now in the research/industry context, spinel specifically for gemstone use is now being grown in Russia.

Before looking at the Russian product, we should be aware that one example of a flux-grown spinel crystal could cause a good deal of alarm among crystal collectors: this is the colourless variety, which could be mistaken for an octahedron of diamond. Similar crystals of colourless corundum, also grown by the flux-melt method, could also be deceptive. If either is encountered, look for the absence of mineral inclusions so frequent in diamond, and for the orientation of triangular surface markings on the faces of the octahedron. In diamond these markings, known as trigons, do not follow the edges of the face, so that they do not point to the apex. In the synthetic corundum and spinel crystals they are reversed and point to the apex. Any colourless or coloured well-formed octahedron with a clear interior should be closely examined, especially when the colour is not familiar: dopants have enabled the crystal grower to produce very beautiful colours; groups of coloured spinel crystals cannot be natural.

When the extra alumina, normally 2.5 times the amount found in nature, is increased to approximately 5 times, the result is a boule which can be cut into a very convincing imitation of moonstone. This arises because a spinel of this composition is unstable, allowing some of the alumina to precipitate out to form crystals from which reflected light returns to the observer as adularescence or schiller, two of the terms used to describe the characteristic glow of moonstone. Moonstone has an SG of 2.57, much lower than that of synthetic spinel at 3.64, and its RI of 1.54 compares with that of 1.728 for synthetic spinel.

Reports of flux-grown red and blue spinel from the Russian Academy of Sciences, Novosibirsk, were circulating in the early 1990s, and I have been able to examine some specimen crystals of both colours. The crystals were well shaped with octahedral form, each one having signs of an attachment point to the growth vessel; a similar feature was also noted from crystals examined by the GIA and reported in the summer 1993 issue of *Gems & Gemology*. The GIA also looked at 9 red and 12 blue faceted stones ranging in weight from 0.19 to 8.58 ct. In the same study were two (rough and faceted) blue flux-grown Russian spinels known to have a higher content of iron than the other samples.

The colour of the red stones was a vivid medium dark and slightly purplish red, some specimens possessing a very slight orange to brown component in incandescent light, this component not showing under fluorescent light in which the stones had an enhanced purplish element. The blue specimens showed a saturated medium dark to dark blue with a slight tinge of violet. Red flashes could be seen when the stones were moved in incandescent light and a grey component was noticeable in fluorescent light.

The RI of all the specimens examined was in the range for natural spinel, while the specimen with the higher iron content gave 1.717. The SG was also in the natural spinel range, all the samples descending easily in di-iodomethane, which has an SG of 3.32. Under LWUV the red stones gave a strong purplish-red to slightly orange–red reaction with a similar but weaker reaction under SWUV. Under SWUV some chalkiness on the edges between faces could be seen, and these also showed a yellowish-orange in some directions. No phosphorescence could be seen after either kind of irradiation. UV testing does not therefore separate the flux-grown material from the natural.

With the hand spectroscope the red material gave an emission line in the red region between 685 and 680 nm with a broad absorption band between 580 and 510 nm. There was a general absorption from the blue material at about 450 nm to the UV. The so-called 'organ-pipe' group of emission lines sometimes seen in natural red spinel could not be detected in the flux-grown material, only one line being present: though, as described above, they can be seen in some of the Verneuil red spinels. The blue specimens absorbed strongly between 635 and 615 nm, between 590 and 560 nm and between 550 and 535 nm. These are characteristic of cobalt, and it is clear that the blue spinels are doped with this element: confirmation is given by their red to orange colour showing through the Chelsea filter. Both red and blue stones also gave a red to orange–red transmission luminescence when viewed in a strong visible light. The red to orange–red colour seen through the Chelsea filter was until recent years a diagnostic sign of a cobalt-doped Verneuil synthetic blue spinel but some natural blue cobalt-bearing spinels have been found in Sri Lanka: in these stones the gemmologist has to rely on the slightly muted orange through the Chelsea filter and more on the natural inclusions present.

Inclusions in the Russian spinels showed clear signs of flux growth, with flux residues forming net-like patterns and jagged edged particles. Some of the flux inclusions contained gas bubbles, while others reflected greyish silver and may have been from the platinum or iridium crucible in which growth took place. In some of the specimens the larger flux inclusions formed pyramidal shaped phantoms in very close alignment with the edges of the octahedron. The red crystals did not show the triangular markings often seen on the faces of natural spinel, but under 80× magnification showed growth hillocks. In the flux specimens dendritic inclusions were observed by the GIA: they gave a metallic appearance in reflected light. No colour zoning could be seen in any of the samples.

Chemical analysis found that titanium – present in natural red spinel – was absent from the synthetic material.

The red spinels can be conclusively identified only by the presence of flux residues or metallic inclusions or, of course, by the absence of natural mineral inclusions. The blue stones show the cobalt absorption spectrum more strongly than the rare cobalt-bearing natural spinels, contain flux residues similar to those shown by the red stones and under LWUV glow with a weak to moderate, somewhat chalky red to reddish purple fluorescence: the same colour, but stronger, could be seen under SWUV. Verneuil blue spinels do not show this colour but appear a mottled blue to bluish white. Natural blue spinels are inert to SWUV.

If and when other colours of spinel are grown, careful testing, probably involving chemical analysis, may be found necessary. The Verneuil product is likely to remain the most common, except for red specimens.

One other use of spinel is the lapis lazuli imitation made by growing Verneuil cobalt-bearing blue material and adding flecks of gold to simulate the pyrite found in much natural lapis. We shall meet this material when we look more closely at lapis lazuli.

Occasionally a pink synthetic spinel may cause confusion but as long as the possibility of such a stone turning up is borne in mind, no difficulty should arise. Flux-grown pinks are always likely to be grown in small quantities.

Lapis lazuli

To the mineralogist the opaque blue material lapis lazuli is a mixture of several different mineral species, most of them blue, which combine to make a rock. To the gemmologist and jeweller lapis is one of the most important ornamental gem materials and one which is often imitated. As we shall see, there is a so-called synthetic lapis too, but this material has some differences from its would-be natural counterpart.

Artefacts made from lapis are often quite large: carvings abound, so an RI test is often impossible due to the lack of a suitable flat surface to be tested. Lapis gives no useful absorption spectrum so the gemmologist is forced back to the microscope and to an SG test, tedious though this may be. Fortunately, there are not too many imitations, and the commoner ones can be tested more easily than lapis itself.

Lapis at its finest is an even dark blue with no trace of any other colour and no whitish patches: these are often seen in the cheaper varieties and are in fact crystals of calcite. It would not be worthwhile imitating or synthesizing cheaper material, so if the calcite patches are seen the specimen will usually be lapis. Lapis may also contain visible crystals of brassy yellow pyrite: these are not unattractive and have been simulated by gold (fool's pyrite!) in at least one of the important lapis imitations.

The commonest of the imitations is dyed jasper whose SG is 2.58 compared to 2.7–2.9 for lapis. Stained with ferrous ferrocyanide this is not unsuccessful as an imitation since quartz flakes may be mistaken for pyrite; blue aventurine glass coloured by copper spangles has also been used. This will also have a lower SG than lapis, and the spangles will easily be seen under magnification.

One of the most popular lapis imitations, at least to the gemmologist, is a variety of synthetic spinel reported to have been introduced to the markets in 1954. The Verneuil-grown spinel was doped with cobalt and heated to at least 2000°C and at one time was known as 'synthetic sintered spinel': specimens were marketed by Degussa of Frankfurt, Germany. When suspected, the higher SG of the spinel at 3.64 compared to the 2.83 of lapis will make artefacts feel heavier than expected: such a test will only be useful to gemmologists and jewellers who are accustomed to handling lapis, however. If a Chelsea filter is handy, the cobalt-doped spinel will show bright red through it while lapis shows a dull brownish-red. Small flecks of gold have been added, as described above, and under magnification these can be seen to be embedded in a material with a granular texture. Since much of this material is fashioned for objects with flat surfaces, such as cuff-links, an RI test may be possible, and will give a vague reading of 1.725 compared to the also vague reading given by lapis near 1.50. Strong cobalt absorption bands in the yellow, green and blue regions will be seen with the spectroscope; lapis gives no useful absorption spectrum. When a strong beam of light from a fibre-optic lamp is passed through a specimen, a distinctive reddish-purple colour will be seen.

Some lapis specimens have been plastic-impregnated to deepen the colour: this has sometimes been carried out on lapis from Chile in which calcite inclusions are often too prominent for successful marketing. When touched with the thermal reaction tester a pungent smell associated with plastics will be given off.

In the 1970s, Pierre Gilson brought out a 'synthetic lapis' in a number of different varieties: some contained added pyrite, but the material is porous compared to natural lapis and has a lower SG of approximately 2.46. When a piece was drawn across an unglazed porcelain test plate it left behind a strongly coloured blue powder: this test (the streak test) with true lapis powder gives a much weaker blue. While the test was being carried out the Gilson material gave a distinctly sulphurous smell: this can be detected in natural lapis but only when a much harsher abrasion is carried out such as a lapidary would give. The Gilson material reacts more strongly than natural lapis when acids are applied. Given Gilson's background in ceramics it is probable that the lapis is made by some similar technique. The material has been found to consist of ultramarine and hydrous zinc phosphates.

As might be expected, natural lapis is sometimes dyed and/or waxed to improve the colour and to minimize the effect of calcite where present. Concentration of dye or wax in cavities and cracks can be seen under magnification, and colour can often be removed with an acetone-soaked cotton swab, (nail-polish remover will also serve). Oiling and waxing with coloured materials is known.

Occasionally a natural mineral may be dyed to resemble lapis: howlite, which has an SG of 2.45–2.58, was reported as a dyed lapis imitation which gave an intense orange fluorescence in a part of the specimen from which dye had been removed. Such a response to UV is reported from untreated howlite from California. Dyed dolomite has also been reported: as a carbonate such a specimen would react with acids. It has an SG of 2.85 and an RI of 1.50 and 1.68. With an easy cleavage, it would be surprising if dolomite is often used in this role.

Turquoise

I may have said elsewhere that amber and turquoise are not the gemstones in which to specialize if you want to make a profit from a gem-testing laboratory! Turquoise can show a wide variety of colours, though always shades of green and blue: it can easily be imitated and it has also been synthesized, or at least, a 'synthetic' product has been advertised.

Turquoise is a phosphate of copper and aluminium and occurs most commonly as microcrystalline aggregates: though single crystals are found in a few places, they are far too small for fashioning. Turquoise is often found with included veining of its host rock and this conjunction can be a pleasing effect to be used by the craftsman setting the stone. Turquoise has a hardness of 5–6 on Mohs' scale and an SG in the range 2.6–2.9 for most specimens. The RI is usually seen on the refractometer as a single shadow-edge at 1.62, but testing in this way is not recommended for porous turquoise as the colour may be damaged.

When examining a suite of turquoise jewellery a close matching of stone colours is certainly a matter for suspicion. While it is possible that such stones are surface-treated turquoise or perhaps the synthetic material, it is much more likely that they are not turquoise at all but glass or dyed chalcedony.

The synthetic turquoise was placed on the market by Gilson in 1972. Like the Gilson lapis lazuli, it appears to have been made using ceramic techniques with grinding, precipitation and/or pressing. It is not made from ground-up natural turquoise since this would introduce iron, which has not been found in the stones. Several different varieties have been made, some with and some without veining.

When the Gilson turquoise is examined, profuse angular dark blue particles can be seen on the surface, set against a whitish ground-mass. The Gilson stones have an SG of about 2.74 and an RI of 1.60. The copper absorption band at 432 nm is not easy to see, but it is hard to see in natural turquoise too, unless it has been colour enhanced or stabilized. This is always one of the hardest absorption bands to see, and patience is needed on the part of the observer. Magnification at 30–40× is the best method of testing.

Gilson made at least two turquoise varieties, one with and one without matrix. Furthermore, there are two distinct compositions, (unaffected by the presence or absence of matrix). One is turquoise with one or two additional phases while the other is a substitute consisting mainly of calcite. Both these materials show absorption bands not seen in natural turquoise. The medium-blue Gilson material has been named 'Cleopatra' and the darker-blue product 'Farah'.

A detailed survey of the Gilson product is by Williams and Nassau in *Gems & Gemology*, Vol. 15, pp. 226–232. The paper also describes other turquoise simulants: these are a simulated turquoise manufactured by the Syntho Gem Company of Reseda, California, 'reconstituted' turquoise made by Adco Products of Buena Park, California, and 'Turquite' from Turquite Minerals of Denning, New Mexico.

Turquite was found to be poor in aluminium but rich in sulphur, silicon and calcium. The Adco and Syntho products were similar to the Gilson material. Structurally only the Gilson product was the same as natural turquoise.

Turquoise imitations can be tested by placing a drop of Thoulet's solution (iodides of potassium and mercury) on the surface, which will turn brown at that point. A white spot in this test indicates natural turquoise. Plastic imitations will have an SG well below 2.62, and often show thread-like markings.

Among the substitutes for turquoise, dyed blue magnesite is quite convincing. It has an SG of approximately 3.0 and an RI of 1.51 and 1.70. As a carbonate, magnesite will effervesce when touched with a drop of warm dilute hydrochloric acid. Dyed howlite has an SG of 2.53–2.59 and an RI of 1.59. Ivory is occasionally dyed with the aim of imitating turquoise, but should always show lines of Retzius (see the section on ivory in Chapter 12): at 1.80 ivory has a much lower SG than turquoise.

Other turquoise imitations include 'Viennese turquoise', a precipitate of aluminium phosphate consolidated by pressing and stained blue by copper oleate, and 'Neolith', a German-made mixture of copper phosphate and the mineral bayerite, an aluminium hydroxide. A drop of hydrochloric acid will give a yellow colour where it touches the surface of either of these products.

Much turquoise when recovered is pale and powdery, the porous structure when pronounced leading to so much light scattering that the stones hardly appear blue. Filling the pores with plastic, oil or wax (all with a higher RI than air) diminishes the scattering, and a darker blue is seen. Nassau (1994, *Gemstone Enhancement*, Butterworth-Heinemann) makes the point that the 'fading' of turquoise is often caused by the drying up of moisture from the pores. Acids from the skin of the wearer of turquoise jewellery can have a similar effect, and sometimes alter the colour from blue to green.

Paraffin wax is a popular impregnating agent, the stone first being dried then soaked in warm melted wax, the process taking several days. Wax is less durable as a filling than polymers of the epoxy type. Immersion in a sodium silicate (waterglass) solution followed by concentrated hydrochloric acid forms a silica gel within the pores, and a colouring agent can be added. Colloidal dispersions of silica in water have also been used.

The methods just described usually do not involve a colouring agent and simple dyeing is usually combined with some kind of impregnation with the aim of consolidating the specimen. Various copper compounds have been used as well as Prussian blue, used to dye chalcedony, as we have seen. Nassau (1994) cites a report which states that because many Egyptians prefer green to blue turquoise, different treatments are used to alter blue material found there to a green colour.

Turquoise can be coated either to protect a layer of dye or to improve a dull-appearing specimen. In one method the surface is etched with acid and a blue epoxy resin applied, the excess being polished away. In another method cited by Nassau (1994) the surface of turquoise beads was painted blue, black paint added to imitate matrix and the whole bead finally coated with clear lacquer.

As always, the microscope has to be used once treatment is suspected. Colour may concentrate in cracks or in drill holes, and some dye may be removable with ammonia on a cotton swab. The thermal reaction tester will cause some coatings to melt locally, and plastic coatings will give the characteristic acrid odour under such treatment. Stabilization of turquoise without the addition of colour is not considered to need disclosure.

The jade minerals

Only the minerals nephrite and jadeite may be called 'jade', and as might be expected both have a large number of imitations: the emerald-green translucent jadeite, with no hint of white streaks, has traditionally been known as 'Imperial jade', and will always be the target of those attempting to imitate or even to synthesize jade.

Jadeite of jewellery quality was reported to have been synthesized by R.C. DeVries and J.F. Fleischer of the Inorganic Materials Laboratory of General Electric in 1984. The manufacturing process involved high pressures, and it was claimed that a number of different colours could be made: the starting material was reported to be glass obtained by melting sodium carbonate, alumina and silica to give the jadeite composition of sodium aluminium silicate. For fine green jadeite chromium would have to be added. Growth took place in platinum crucibles in air at a temperature of approximately 1550°C. The molten liquid was cooled, the glass then being crushed and refined until a homogeneous composition was achieved. Since the liquid tried in the first experiments remained viscous even at 1550°C, gels were used for later production.

Since 1984 there have been no reported inroads of this material into the gem market, and we have to conclude that commercial considerations have so far ruled it out as a serious contender.

Both jadeite and nephrite are composed of minute interlocking crystals whose fibrous nature gives both minerals great toughness even though neither has a hardness of more than 7 on Mohs' scale. More importantly for the gemmologist, the fibrous structure makes the jade minerals reasonably easy to dye in a range of colours. Natural nephrite usually shows more restrained colours than jadeite: both have a range from whitish through yellow and green to brown and black. Lavender jadeite is also found.

Treatment may be to enhance or lighten the natural colour, and heating is well known to lighten dark green specimens. If the material contains inclusions of iron compounds, their yellow to brown colours may be heated to give brownish or reddish colours. Whether or not this is to be done can be decided at an early stage after recovery since jadeite is often covered by a skin in which iron is present. If jade has already been dyed the process of heating will destroy any colour thus gained: lavender jadeite is particularly heat-sensitive and may lose its colour at temperatures as low as 220–400°C (Nassau, 1994), though many examples keep their colour at least to 1000°C.

Irradiation is not regularly used with the jade minerals, though reports appear from time to time. Dyeing is by far the commonest treatment: with a more porous structure, jadeite is easier to dye than nephrite and is more translucent. Enhancement of pale material, quite common in jadeite, usually takes the form of some shade of green, although other colours are now appearing on the market: among them are lavender and violet. Carved material may be treated in such a way as to enhance details by the judicious use of brown staining.

Some of the dyes used are stable to light while others, including many aniline dyes, are not. The number of possible dyes is great, and many reports have been published. Short of uncertain fade tests, the usual way to detect dyed material is to look, under magnification, for signs of dye concentration in out of the way places on the specimen. These will look notably darker than surrounding areas. Places such as drill-holes and in the intricacies of carving should be closely examined.

Jadeite sometimes needs to be bleached to get rid of the brown stains which are very common. Soaking in acids will normally remove the stains, and other colours are not usually affected. One technique is to follow the acid treatment by rinsing in water and neutralization with an alkaline substance, leaving the jade pale: increased porosity then makes it easier for wax or polymers to be introduced. These may be coloured (usually green) or the specimen may be dyed first so that the filling can then be colourless. The name 'B-jade' has been used for some of this material; 'C-jade' is also used for polymer-impregnated green jadeite of lower quality. If a specimen is suspected, the presence of brown stains should not allay suspicion since a few brown stains are sometimes deliberately left behind!

This is fairly robust treatment, and some specimens may be weakened: the colour may darken and a yellow exudation from the acid may become apparent.

Wax may be used to hide small fractures near the surface: paraffin wax is easily obtained and is often used on carvings and cabochons. Nassau (1994) makes the point that flaws concealed in this way may escape the notice of the setter, who may then unwittingly damage the specimen. Impregnation and coating by polymers has been used to improve transparency, and foiling is still carried out. In Hong Kong, concave mirrors with small openings may be used as the back of a setting for jade, according to Crowningshield via Nassau (1994). Imperial jade has been simulated by placing a thin dark-green polymer layer over translucent white jadeite: a much more dangerous imitation of Imperial jade is the ingenious hollow cabochon.

A pale piece of jadeite is hollowed out to give the shape of a cabochon and another piece fashioned to fit the hollow base. The two are cemented together, with green dye coating the inside of the hollow. On completion the stone appears to be solid, although many cutters leave a ridge where the base joins on. From a casual inspection the stone appears to be solid and the colour is good and even: the thin walls of the cabochon give something of the translucency expected from jadeite of high quality, even Imperial jade. The spectroscope is probably the best instrument to use on jadeite or possible jadeite: green material should give both the characteristic chromium absorption spectrum and a prominent band at 437 nm, this showing jadeite as a member of the pyroxene family of minerals. When the colour is the result of dyeing, a woolly rather than sharp absorption band is seen in the red region instead of the chromium absorption features: this effect seems always to be present in stones owing their colour to dyestuffs. A few naturally coloured dark-green jadeite specimens will show a broad absorption in the red portion of the spectrum.

When jadeite has been bleached by acid treatment the SG may be lowered slightly to about 3.32; the material used for impregnation may contain gas bubbles and the specimen may contain fractures. Any kind of bleaching is hard to detect with certainty, and laboratory testing may be needed. Glass imitations will normally feel less heavy and contain gas bubbles which, even if the material is opaque, may show close to the surface.

A useful general point is that through the Chelsea filter natural green jadeite remains green whereas many of the dyed imitations show red or pink. It should often be possible to carry out an SG test even if the specimen is quite large (and, of course, unmounted): all that is needed is a vessel large enough to accommodate it. When a specimen has a flat and accessible surface the refractometer can be used. The SG of jadeite is close to 3.33 and that of nephrite approximately 3.0: RI values are 1.66 and 1.60, respectively. To avoid shock, remember that even though you might expect birefringence to be undetectable in a crypto-crystalline structure, a specimen of jadeite with a flat back will in fact show two shadow edges, at 1.654 and 1.667, on the refractometer.

The glass jade imitation Meta-jade or Victoria stone is described under glass.

The synthetic garnets

The name synthetic garnet is used by crystal growers and materials scientists to denote a group of oxides of the cubic crystal system with the general composition $A_3B_5O_{12}$ in which A and B represent a number of possible elements, and O is oxygen. More than one representative of this group has spent some time as the 'best diamond simulant', as group members are hard and transparent and, as members of the cubic crystal system, singly refractive so that there will be no doubling of back facet edges seen when viewed through the table facet.

Gemmologists should remember that the term 'synthetic garnet' will spring out at them as soon as they desert some of their long-established textbooks and turn to some of the literature of crystal growth: to the mineralogist the garnet minerals form a group whose members are all silicates rather than oxides, and gemmologists have to take their line from the senior earth sciences and from chemistry. Gemmologists and students are unlikely to confuse the two groups of species 'in the field', since there are many obvious differences. Confusion could very easily arise, however, when the synthetic garnets are colourless and simulate diamond, and when they are doped with foreign elements to simulate coloured diamonds and other gem species. It does not help the gemmologist or the jeweller that most traditional tests are not easy to apply to these materials, but, as so often happens, when one test fails another is ready to take its place!

We have briefly looked at some examples of the synthetic garnets when considering diamond and its simulants, but many of them have lives of their own outside the world of diamonds. The crystals are grown by the flux-melt method or by crystal pulling, both methods yielding quite large specimens so that the potential for large bright hard colourless or coloured stones is high.

The best known of the synthetic garnets is commonly known as YAG, yttrium aluminium garnet. As a diamond simulant, YAG held the top position from its first appearance on the market until the coming of cubic zirconia (CZ), which is more lively in appearance. YAG has a dispersion of 0.028 compared to the 0.044 of diamond and the 0.059–0.065 (depending on composition) of CZ. YAG has a hardness of about 8.5 and an SG of 4.55 compared to the 3.52 of diamond. The RI is 1.83 (diamond is 2.42). The superior hardness and RI of diamond makes it virtually impossible for simulants to succeed.

Many trade names have been used for YAG, 'Diamonair' being one of the most popular ones, though 'YAG' is the common name among gemmologists and jewellers.

In conditions where a quick test appears to be necessary (the trade should never allow itself to be pushed into a quick test for any gemstone!) the tilt test might be useful, providing that a single source of light is available (a desk lamp will do). If a faceted diamond is held over a dark background with the observer looking down upon the table facet and this facet is gradually tilted in a direction away from the observer, diamond simulants will appear to lose light the further the angle of tilt: this does not happen with a well-proportioned brilliant-cut diamond. Both YAG and CZ, with RI values lower than that of diamond, will lose light in this way.

Read (1991) describes an interesting 'dot–ring' test in which the polished specimen is placed table facet down upon a small black dot drawn on a piece of white paper. In the case of diamond simulants the dot will appear as a ring surrounding the culet. This will not happen with diamond. For all simple tests the diamond should be well proportioned.

More confidence should be placed on the differing response of YAG to the reflectivity meter, which is described elsewhere (see Chapter 5). The high hardness and RI of diamond serve to produce a uniquely reflective surface which allows the reflectivity meter to distinguish it easily from YAG. Read (1991) also explains how checking the weight of a polished diamond or diamond simulant against its girdle diameter gives a rough estimate of SG stones tested in this way should have ideal or near-ideal proportions. For a stone with girdle diameter of 2.0 mm the carat weight of diamond will be near 0.03 ct, that of CZ 0.05 ct, of YAG 0.04 ct and of GGG – gadolinium gallium garnet – 0.06 ct. For a girdle diameter of 6.5 mm the corresponding weights will be 1.0 ct, 1.65 ct, 1.30 ct and 2.0 ct.

Where YAG and the other synthetic garnets perhaps give most danger is when they have been doped to give bright and sometimes unusual colours. The rare earths, a group of elements forming neighbours on the periodic table, are routinely chosen as dopants, not so much with ornamental use in view but to achieve particular industrial or research ends. As it happens, some of the rare earths which are added in their oxide forms to the starting materials of the synthetic garnets before growth commences give bright body colour to the grown crystals and to gemstones which are fashioned from them. A number of rare earth dopants give fine-line absorption spectra with a great number of lines extending across the spectrum as seen with the hand spectroscope: such absorption spectra are never seen at such strength in natural minerals, nor, fortunately, in the natural gemstones for which it may be hoped that the doped crystals will be mistaken.

Some of the rare earth dopants, their colours and the properties of gemstones which may appear in commerce are given by O'Donoghue (1983). Yttrium gives a green colour and stones have been recorded with an SG of 4.60 and an RI of 1.834. Similar data are given for stones doped with terbium (pale yellow), 6.06 and 1.873; dysprosium (yellow–green), 6.20 and 1.85; holmium (golden yellow), 6.30 and 1.863; erbium (yellowish–pink), 6.43 and 1.853; thulium (pale green), 6.48 and 1.854; ytterbium (pale yellow), 6.62 and 1.848; lutecium (pale yellow), 6.69 and 1.842. Blue colours, also involving a multiline absorption spectrum, appear in the trade from time to time, and are probably coloured by cobalt by a process which may also involve silicon. Some green YAG was at first thought to be fine demantoid garnet, though the resemblance is not particularly close: however, an examination of the absorption spectrum shows characteristic rare earth elements as well as those associated with chromium, which is also present; manganese has been used to produce darkish red material, and yellow can be made by doping with titanium.

Some green YAG shows red through the Chelsea filter, and some colourless material has been found to fluoresce yellowish under UV and mauve under X-rays. These effects are not universal, and their presence or absence should not be taken as diagnostic features: it is probable that they are caused by impurities which may be present in some batches and not in others. Nor is the fluorescence spectrum particularly useful.

The reflectivity meter will show that YAG cannot be diamond, but as most of these instruments are made solely to state 'diamond/not diamond' without disclosing the identity of the latter, gemmologists who really need to know what a specimen is, in the absence of an absorption spectrum in colourless YAG, will have to look for inclusions of angular flux particles which will appear only in flux-melt material and will be absent from pulled crystals.

My own experience with coloured YAG (this applies also to CZ) is that a colour-change variety is probably the commonest after colourless material. This material appears a lilac or a pink colour depending on the type of light obtaining at the time of viewing. The cause of colour is the rare earth neodymium, whose absorption spectrum is easily seen and interpreted. Colour change is not especially rare in the mineral world, though gemmologists are inclined to treat the effect with reverence!

How much YAG will continue to appear on the gemstone market depends upon demand: it is not expensive to produce and although it was originally grown for laser work (pulled crystals) and for research into magnetic and insulating properties (flux-grown crystals), a certain amount is grown specifically for faceting. Gemmologists will have to take care with brightly coloured faceted stones with a high lustre. As so often with coloured stones, the spectroscope will be the instrument to choose first. We should always remember that fashion is important with gemstones: though CZ is now the 'top synthetic', YAG has not gone away!

Though YAG and its analogues (materials with the same structure but with varying compositions) form an important group of man-made gemstones, many are quite hard to grow and expensive to dope. One fairly expensive species has had some success both as a diamond simulant and as a coloured stone: this is gadolinium gallium garnet (GGG – pronounced 'three G's'), with the composition $Gd_3Ga_5O_{12}$. This has a much higher SG than YAG, at just over 7 and a large stone of GGG would feel heavy when set in a ring.

GGG began life as a research material used for magnetic bubble domain memory units. For this a very high degree of purity is needed. GGG has a hardness of 7, low for a serious diamond simulant, and an SG of 7.02. The RI is 1.97 and the dispersion 0.045. GGG, as a synthetic garnet, belongs to the cubic crystal system, and thus shows no birefringence.

Both gadolinium and gallium are expensive compared to the constituents of YAG, but some colourless and some doped specimens are about in the trade. Since the coloured GGG is doped with rare earths, the absorption spectra are equally distinctive and diagnostic. As with some YAG, impurity elements may give a brownish tinge to colourless specimens: even the UV present in sunlight has been known to produce this effect. Different fluorescent behaviour may be seen but is not especially useful in testing.

Two final points on the synthetic garnets: one concerns a name often found when rare earths are described. 'Didymium' is not an element but a short substitute for 'neodymium + praseodymium': gemmologists who also collect mineral crystals should beware of (or look out for) crystals of the synthetic garnets. Some of the doped ones are very beautiful and while they should not deceive experienced mineralogists they do show garnet forms of rhombic dodecahedra combined with icositetrahedra.

Cubic zirconia

Why cubic? Why zirconia? To answer the second question first, the suffix 'a' to denote the oxide of an element is a convention, thus silica, alumina and so on. Zirconia is, then, zirconium oxide with the formula ZrO_2. A mineral with this composition exists in nature but has no ornamental significance, and is in any case a member of the monoclinic crystal system: it is not, therefore, *cubic* zirconia.

What is so special about a possible cubic zirconium oxide that it is worth growing on a large scale? If a material can be cheaply produced and it is hard, transparent and shows no birefringence (which would apparently double back facet edges and inclusions when a faceted stone is examined through the table) there is always the possibility that it will make an effective diamond substitute. In addition, CZ is easier to grow than the synthetic garnets since no crucible is needed: all that the grower has to do is to find a growth method which can cope with the very high melting point of CZ of 2750°C. Though a cubic modification of zirconia was discovered in 1937, successful synthesis did not take place until the 1970s, when it was finally achieved in Russia by scientists using a new technique of skull melting.

Even though a growth method was in place, monoclinic zirconia had to be avoided, and this was done by adding a stabilizer, either calcium or yttrium, borrowing techniques used in the production of hard and heat-resistant ceramics.

CZ is easily the best imitation of diamond on the market today: compared to diamond it has a hardness of between 8 and 8.5 and an SG of about 6.0, much higher than the 3.52 of diamond. The RI is close to 2.16 compared to 2.42 for diamond; its dispersion is 0.060 while that of diamond is 0.044. Gemmology students will appreciate a formula which I found in the Dick Francis novel *Straight* (Michael Joseph, 1989): CZ = C × 1.7. For non-gemmologists, C is diamond! SG is the context.

As a crucible is not used for the growth of CZ there can be no useful metallic fragments turning up as inclusions. For so important a material there is little to help the gemmologist without recourse to the reflectivity meter, many models of which will only register diamond/not diamond, not saying what a non-diamond actually is. CZ can be coloured, of course, by the use of rare earths and other dopants. In many cases the dopant will show a characteristic and diagnostic absorption spectrum which will help, though such a test will not distinguish between doped CZ and doped synthetic garnets. Diamond will show a range of natural inclusions (in most cases), and it is well worth getting a mental picture of them.

Colourless CZ occasionally shows a greenish-yellow or even reddish fluorescence but this should not be expected as a matter of course: it seems to be associated with yttrium-stabilized material. Sometimes rows of small semi-transparent isometric crystal-like cavities can be seen in the yttrium-stabilized stones: these may merge into hazy stripes of tiny particles. Calcium-stabilized zirconia may show a distinct yellow fluorescence, but is normally free from inclusions.

The gemmologist without instruments will note that liquid will form a coherent drop on the surface of diamond whereas the CZ surface will cause a drop to break up into beads. No doubt special liquid and applicators are available, but I would expect anyone dealing regularly with gemstones to have the reflectivity meter at their side. The coherent drop is a characteristic of diamond which is shown by none of its simulants.

Among the dopants are cerium oxide to give red, orange and yellow colours, europium and holmium oxides to give shades of pink, chromium, terbium and vanadium oxides for shades of green and copper, and manganese and neodymium oxides for lilac and violet colours. Neodymium-doped stones give the lilac-to-pink colour change. Though reports of ruby-red and emerald-green CZ appear regularly, it is not certain at the time of writing whether or not these colours are in fact achieved. Sapphire-blue stones have also been reported, and it is said that emerald and blue sapphire colours can be produced only when the proportion of stabilizer is much higher than usual. The name C-Ox was used for Russian CZ in the blue and green colours.

Not all CZ is transparent, as white, pink and black translucent to opaque material has been manufactured in Russia. Sold as beads or cabochons, the colours are a milky white and a uniform medium pink, showing banded or striated colour distribution when viewed by strong transmitted light. Some specimens have been found to show dark brownish-red under strong transmitted light but all specimens have a notably high lustre. When suspected, such pieces would need to be tested with a reflectivity meter if identification was uncertain or vital: however, the species imitated in this way would not be very expensive ones.

A dark yellowish-green specimen of CZ showed red through the Chelsea filter and looked like green tourmaline of good quality: absorption bands at 607, 583, 483, 472 and 450–443 nm showed that the specimen was not tourmaline.

The growth method used to produce CZ results in quite characteristic columnar crystals with no crystal forms visible: thus it is not possible for CZ crystals to be mistaken for natural ones of some other species, unless someone takes the time and trouble to cut them for this purpose. None the less there have been occasional reports of octahedra with artificially etched trigons on their faces which will have been cut to this shape. A final useful test for a faceted CZ is to place it table down over a strong light source: the pavilion facets will often show red to yellow colours.

Topaz

While the pink variety of topaz may occur naturally (as in Pakistan and Brazil), a pink colour can be induced in some reddish-brown crystals, provided that some chromium is present. The colour is permanent and thus needs no disclosure: in any case there is no way in which artificially coloured pink and natural pink stones can be distinguished from one another.

Blue topaz, which could be mistaken for aquamarine, is easily distinguished by standard tests. The darker-blue stones now on the market have been artificially coloured by irradiation but, apart from the colour, signs of the treatment are not apparent. Gamma-ray treatment followed by heating has been used to deepen the colour of pale-blue topaz, but placing stones in a linear accelerator ('Linac') is preferred since higher available energies can give darker colours: heating still forms the second part of the treatment. In the early days of blue topaz treatment a few stones showed residual radioactivity, but this should not be expected with consignments today, and in any case the radioactivity quickly diminished.

Blue topaz may also be darkened by placing it in a nuclear reactor, after which treatment heating is not always needed. The trade name London blue is often given to stones with a dark blue colour with some inkiness: heating can give a lighter and less inky appearance. Stones treated in a reactor will be radioactive for 1–2 years, and gem-producing countries have legislation controlling their release on to the markets. Recently, some blue topaz (Super blue, Swiss blue, American blue) has appeared on the market. The stones are believed to have been reactor-treated, Linac treated and finally heated to diminish inkiness. Aqua Aura topaz is a name given to topaz crystals on which a thin film of gold has been deposited. As in the Aqua Aura quartz, crystals show a blue colour with surface iridescence.

Reports of interesting and unusual examples from the literature

Items in this section have been chosen to illustrate points made in the chapter and to bring one-off items to your notice.

A Russian synthetic alexandrite is reported in the winter 1995 issue of *Gems & Gemology*. Schmetzer examined around 200 crystals of the alexandrite, manufactured in Novosibirsk and sold in Bangkok. Any mention of Bangkok immediately informs the gemstone trade that specimens are on sale, so a watch should be kept for cut stones showing prominent growth zoning as closely packed parallel lines intersecting at angles. Some specimens showed an intense red core with a lighter red rim: between the two was an even more intense red boundary. Any alexandrite offered with an intense red in any part of the stone should be regarded with suspicion, as the natural stone rarely shows such a colour.

For the gemmologist who has a chance to examine the crystals, characteristic hexagonal outlines are reported from about 90 per cent of specimens examined.

A fracture-filled alexandrite was identified by the GIA and reported in the Fall 1995 issue of *Gems & Gemology.* The specimen weighed 0.45 ct and was a semi-transparent pear-shaped modified brilliant with a colour change from dark bluish green in daylight-equivalent fluorescent lighting to a dark reddish purple in incandescent light. It was shown to be a natural alexandrite by normal gemmological testing although it showed a rather high RI of 1.753–1.761. Energy-dispersive X-ray fluorescence (EDXRF) analysis showed unusually high levels of chromium, titaniun and iron. Several surface-reaching fractures were observed, these containing a transparent colourless material. When the tip of a thermal reaction tester (hotpoint) was brought near to the surface the filler was seen to flow within the fractures.

A flux-grown alexandrite showing no diagnostic inclusions is reported by the GIA in the fall 1995 issue of *Gems & Gemology.* The stone, weighing 1.08 ct, was a transparent oval brilliant with constants and colour change identifying it as alexandrite. No routine tests could diagnose a natural or artificial origin. In this case infra-red spectroscopy showed that water-related absorptions around 3000 cm^{-1} were absent – they are normally seen in natural alexandrite. It was unusual to find traces of residual flux.

An alexandrite weighing 3.59 ct was examined by the GIA and found to contain numerous thin, short and highly reflective needle-like inclusions as well as abundant dust-like pinpoints. The needles appeared to be mostly straight though some were slightly curved. Curved colour banding could be seen parallel to the girdle plane when the stone was immersed in di-iodomethane, though the curvature was slight and might have appeared owing to some kind of optical distortion. The banding was straight through one end and in the centre of the stone, but curved at the other end. The colour change was from red–purple in incandescent light to both green and purple in fluorescent light. Standard tests showed the stone to be alexandrite but EDXRF testing detected chromium and vnadium in very similar proportions to those found in reference samples of Czochralski-pulled synthetic alexandrite. Infra-red spectroscopy showed that water was not present, a characteristic feature of melt-type synthetics. The stone is described in the spring 1993 issue of *Gems & Gemology.*

The trade name 'Allexite' has been used for a synthetic alexandrite manufactured by The House of Diamonair, whose publicity material identifies the crystals as Czochralski grown. A stone examined by the GIA and reported in the fall 1992 issue of *Gems & Gemology* had a strong colour change from reddish purple in incandescent light to bluish green in daylight or fluorescent light. This change is similar to that shown by the finest Brazilian alexandrites. *In visible light a strong red luminescence (red transmission) was clearly seen. The RI was found to be 1.740–1.749 with a birefringence of 0.009. There was a strong red fluorescence under LWUV, and a moderate red reaction to SWUV. The stone showed red through the Chelsea colour filter, and under magnification distinct curved striae could be seen.*

Filling is not often seen in alexandrite, but one case is reported in the fall 1992 issue of *Gems & Gemology.* The filled cavity was seen on the pavilion of a natural alexandrite, mounted in a ring. The stone measured approximately 9.20 × 5.30 × 3.68 mm. *The filling, in a large negative crystal, could easily be seen when the stone was viewed through the table and a gas bubble proved the nature of the filler. When the thermal reaction tester was brought close to the filled area, the filler softened and gave off a little smoke, suggesting that a polymer had been used.* The glass type of filling so often seen in corundum was not present in this specimen.

A synthetic flux-grown alexandrite submitted to the GIA and described in the fall 1988 issue of *Gems & Gemology* contained *unidentified crystals as well as minute, well-spaced whitish pinpoint inclusions. Traces of flux were not seen. The stone had a good colour change, and a strong red fluorescence under LWUV, as well as an oily appearance that the GIA had found in other synthetic alexandrites. Crystal inclusions are rare in flux-grown material.*

A composite jadeite imitation is reported in the summer 1996 issue of *Gems & Gemology*. The 239.37 ct statuette was made from a calcite and plastic composite, dyed to resemble jadeite. With the refractometer the birefringence blink, indicative of a carbonate, gave the piece away; it also showed an SG of 1.98 (jadeite is 3.33). Under magnification the specimen was found to consist of white grains in a mass of transparent colourless or green material. The resinous lustre and mottled appearance were quite suggestive of jadeite.

A jadeite doublet is described in the fall 1986 issue of *Gems & Gemology*. The specimen appeared a fine green but consisted of a very thin green layer on the top and another, thicker white layer beneath (the thicknesses were 0.1 mm and 2.2–2.3 mm, respectively). *The layers were joined by a slightly yellowish cement containing numerous gas bubbles.* The dark-green upper layer was mottled and contained many nearly colourless veins: chloromelanite was suggested. The white lower layer had a distinct crystalline structure. *RI readings were 1.64–1.74 on the green layer, which showed characteristic chromium absorption lines with the hand spectroscope.* Both layers were identified as jadeite by X-ray diffraction analysis, but the variation in RI of the upper layer remains unexplained.

Jadeite has been divided into three categories, A, B and C, the classes referring to jadeite treated only by accepted surface waxing, jadeite with the subsurface polymer impregnated (the specimen having been first bleached with acid), and jadeite impregnated by a process involving dyeing. A paper in the *Journal of Gemmology*, Vol. 23(7), (1993), describes how the types can be distinguished by the use of a simple test involving a drop of concentrated hydrochloric acid. When placed on a cleaned surface, the acid will be drawn beneath the surface of type A jade by capillary action and an aureole will be seen surrounding the point of application some minutes afterwards. With type B jadeite the acid remains on the surface until it evaporates, since the polymer impregnation has sealed the surface. In type C jadeite the stain markings are shown up by magnification of the acid droplet.

Specimens of bleached wax- and polymer-impregnated jadeite can be distinguished by X-ray photoelectron spectroscopy (XPS), a technique for laboratories only and described with illustrations in the *Journal of Gemmology*, Vol. 24(7), (1995).

A string of beads showing green and white colours and offered as 'moss in snow' jadeite is reported in the winter 1987 issue of *Gems & Gemology*. It consisted of 8 mm quartzite beads selectively dyed. *The RI was found to be 1.55, consistent with quartz, and the beads did not show red through the Chelsea colour filter. Nor did they show the customary 'dyestuff' absorption centred near 650 nm, usually quite easily seen with the hand spectroscope. With the lens, dye could be seen concentrated into cracks, and an acetone-soaked cotton swab easily removed it.*

Nephrite is not too difficult to imitate, and various materials have been used to simulate this valuable jade mineral. In the spring 1987 issue of *Gems & Gemology* a strand of dark green 10 mm beads is described: the beads were stated to be 'imitation nephrite' but were found to be dyed quartzite. *Gemmological testing gave an RI of 1.55 and an SG of approximately 2.65. The structure was seen to be a crystalline aggregate when a bead was magnified, and the presence of a broad absorption band at 650 nm seen with the hand spectroscope proved the presence of a green dye.*

While plastic-impregnated jadeite has been found to give normal constants, *specimens on the whole did not show the fine surface depressions characteristic of jadeite and concentrations of the coating substance were found in irregularities on the surface. With the thermal reaction tester a melting or softening of the coating occurred.*

A carved sphere of 'jade' offered for sale in Hong Kong's jade market and described in the fall 1990 issue of *Gems & Gemology* was opaque to semi-translucent, with a mottled dark reddish-brown colour grading to medium yellowish brown and with around 40 per cent of the area a strongly mottled medium yellow–green to very dark yellowish green. *A birefringence blink was seen on the refractometer, and with the Chelsea filter the greenish areas looked greyish while the brown ones appeared reddish brown. The hardness was near 3, and there was no luminescence observable.* X-ray diffraction showed the material to be largely calcite with some serpentine, the piece having been selectively dyed and coated with paraffin or wax.

While standard gemmological tests are often replaced by such techniques as infra-red spectroscopy in determining whether or not a jade specimen has been impregnated with a polymer, *the gemmologist or jeweller with a keen eye, lens or microscope can detect concentration of the impregnating material in cracks or fractures, or in recessed areas on carved artefacts.* The GIA in the fall 1994 issue of *Gems & Gemology* describe a graduated necklace of mottled green and white beads about 9.50 to 5.90 mm in diameter. *Standard gemmological tests proved the material to be jadeite and that the colour was natural. A colourless polymer layer could be seen in the drill holes of most of the beads, and the presence of the polymer was confirmed by infra-red spectroscopy.*

Impregnation of jadeite is more often seen in green specimens, as this colour is perceived to be the most desirable. In the spring 1994 issue of *Gems & Gemology* a 15.86 ct lavender oval cabochon is identified as jadeite by normal gemmological tests: *under magnification small cavities could be seen, each containing a transparent colourless filling material.* Infra-red spectroscopy showed a strong absorption at about 2900 cm^{-1}, which is characteristic of a synthetic resin.

In the Fall 1993 issue of *Gems & Gemology* a fine green jadeite pendant measuring 39.36 × 17.52 × 6.20 mm is shown to have been bleached and polymer impregnated. The fine colour was quite likely to deceive, but the infra-red spectrum proved the presence of a polymer. *The point of this is that dealers may not even consider that a particular piece has been treated – they are being forced to 'think treatment' for fine pieces at least since nothing visible to the eye gives the treatment away.*

Jadeite may be coated with varnish to improve its appearance and a piece is described in the fall 1994 issue of *Gems & Gemology.* A large jadeite bead of greyish-purple colour was coated with a layer of mottled green varnish. *Some of the coating could be seen to have spalled off, revealing the true colour beneath. The RI at 1.52 was presumed to be that of the coating, as that of jadeite is 1.66; similarly, the SG of the piece, 3.29, was below the normal value for jadeite (3.33). The piece showed a strong chalky blue fluorescence under LWUV with a weaker reaction to SWUV. Natural jadeite normally fluoresces a spotty yellowish-white. On contact with the thermal reaction tester the coating melted.*

A similarly coated jadeite pipe with some of the coating chipped *gave an RI on the coated area of 1.54 and of 1.66 on the exposed area. The 'pyroxene' absorption line at 437 nm showed that the greater part of the piece was jadeite. Two layers of coating were revealed on examination with the microscope: the lower layer was mottled green and the upper one either colourless or a uniform very light yellow – perhaps intended as a protective varnish. The double layering shows up distinctly under UV, where both layers gave a yellow–green response with differing intensities. The underlying jadeite did not fluoresce.* Infra-red spectrometry showed a strong absorption near 2900 cm^{-1} which is not present in natural jadeite but is characteristic of an organic polymer. First recourse to the infra-red spectrum with no other test, however, might wrongly have identified this coated piece as B jade.

Bleached and impregnated jade, known as B jade, is reported in the winter 1994 issue of *Gems & Gemology.* The high-quality bangle was patterned green and white. Acid treatment is sometimes used to remove brown stains which may outline individual jadeite crystals and a honeycomb pattern formed by the grain boundaries may be seen. After this bleaching process, waxing with a neutral-coloured polymer or wax fills the voids left by the removal of the impurities and makes the piece more translucent and more uniformly coloured. While grain boundaries are still visible through the microscope the treatment leaves them less obvious to the unaided eye. One authority states that the 'beehive' structure is one of the chief ways of identifying type B jadeite. The treatment reduces the well-known toughness of jadeite.

The GIA reports on different lighting techniques used with a view to making the honeycomb structure more visible to the gemmologist. Reflected light was preferred to strong illumination by light transmitted by a fibre-optic source. Filled cavities were also noticed: they were flush with the polished surface and their contents burnt slightly when the thermal reaction tester was brought close. This is the characteristic polymer reaction. Further tests with infra-red spectroscopy showed strong absorption between 2800 and 3000 cm^{-1}, which is characteristic of polymer materials. *The bangle floated in di-iodomethane, indicating an SG range of about 3.20–3.25 – this would be expected for a specimen of B jade, but is lower than the figure for untreated jadeite (3.33).*

Most gemstones do not make noises, but this item gave a dull or muffled sound when tapped with steel tongs. Untreated jade gives a distinct ring. In addition, a drop of hydrochloric acid remained intact on the surface rather than 'sweating' around the drop being seen as in untreated jadeite.

A bleached and treated specimen of mottled green jadeite is reported in the spring 1995 issue of *Gems & Gemology.* The cabochon weighed 6.78 ct and gave an RI of 1.66 (consistent with jadeite) by the spot method: also consistent with jadeite was the SG of 3.34 and the presence of chromium lines in the absorption spectrum. While the stone did not respond to SWUV, some areas fluoresced a faint green under LWUV. *Using reflected light the stone showed an etched appearance with the aggregate structure showing as many interlocking grains in various directions, some of the grains being preferentially eroded. This suggested that the specimen had been bleached and polymer impregnated.* This was confirmed by infra-red spectroscopy and the specimen was confirmed as B jade, a trade name for impregnated jadeite of natural green colour.

A double-strand necklace containing both treated and untreated jadeite beads was examined by the GIA and described in the spring 1995 issue of *Gems & Gemology.* While standard gemmological testing proved all the beads to be jadeite, *some in both strands fluoresced a weak to moderate yellow under LWUV while the remainder gave no reaction. It has been found that most polymer-treated jadeite will fluoresce though some untreated material will give a weak yellow fluorescence under LWUV.* It was unusual to find both types of bead in a single piece of jewellery. The nature of both types of bead was proved by infra-red spectroscopy.

Of the two jade minerals, colour improvement by dyeing is more likely to be found in jadeite than in nephrite, However, five pieces of dyed nephrite are reported in the spring 1995 issue of *Gems & Gemology.* The oval mottled green cabochons ranged in weight from 6.38 to 7.12 ct and resembled good-quality jadeite. *Simple gemmological testing gave an RI of 1.61 and an SG 2.95–2.96, all consistent with nephrite. Dye could be seen to have concentrated in cracks, and with the hand spectroscope an absorption band could be seen in the red region, a feature characteristic of dyeing and commonly seen in dyed jadeite and quartzite.*

A glass imitation of jadeite was encountered in Ho Chi Minh city by GIA staff during 1995 and reported in the summer 1995 issue of Gems & Gemology. The material was apple green with an inclination towards yellow, showing a mottled appearance with slightly circular whitish areas which had deeper colour and enhanced transparency around them. *Spherical gas bubbles pervaded the specimen, some appearing as surface cavities which unmistakably indicated glass. The white areas were seen under magnification to be associated with bundles of fibrous inclusions. The effect probably arises from devitrification in the glass – a phenomenon that has traditionally confused gemmologists by its superficial similarity to natural inclusions.* The material known as 'Meta jade' is similar, with angular fibrous patches of lower transparency in a mainly transparent ground-mass. The green material had an RI of 1.51 compared to that of Meta-jade of 1.48, and showed an absorption spectrum with general absorption above 590 nm and below 510 nm (Meta-jade shows general absorption above 560 nm and below 480 nm). Fluorescence was vague, and neither material showed any colour through the Chelsea colour filter. In the Vietnamese material, rubidium, yttrium and zirconium were present: they have not been found in meta-jade.

Dyed green quartz is a popular imitation of jade. In *Gems & Gemology* for summer 1995 the GIA report a dyed green quartzite (metamorphic rock consisting largely of quartz grains) fashioned into an oval cabochon measuring 30.25 × 15.98 × 5.50 mm, which closely resembled fine-quality jadeite. *The RI found by the spot method was 1.55 (jadeite would be 1.66), and magnification showed dye concentration between the grains. Dye was also confirmed by the characteristic absorption band centred at about 650 nm: this is seen in most dyed jade imitation materials.* By the use of infra-red spectroscopy the presence of a substance similar to the synthetic resin Opticon was established. Since the absorptions in the quartzite were much weaker than those previously reported for impregnated jadeite it is probable that much smaller amounts of polymer were used.

Serpentine is one of the most frequently used natural substances to be used as a jade imitation, whether the impression of jadeite or nephrite is sought. While the serpentine minerals are fibrous, they are also fine grained: jadeite and nephrite are formed of interlocking or enmeshed crystals, which make the specimen very tough and much harder to fashion than serpentine. A report in the winter 1995 issue of *Gems & Gemology* cites a serpentine specimen which could have been nephrite on first appearance. The specimen was green and black and partially polished. Routine gemmological tests could have given misleading results as fine-grained minerals can show the properties of individual grains rather than those of the piece as a whole. In this case the refractometer gave an RI near 1.57 (nephrite would give about 1.60), and the SG was determined as 2.63 (nephrite is about 3.00). While the gemmologist will want to know what the specimen is, the commercial world will want to know only 'is it jade or not?'. *The serpentine was shown to be softer and less tough than nephrite by the presence of many fine scratches and of rounded edges on small fractures.* This test can be easily carried out with the 10× lens. *If an ultra-violet source is available (and it can be a very useful counter display), some serpentine will give a rather indistinct mottled chalky-blue fluorescence under LWUV. Neither of the jade minerals will fluoresce at all.* Black inclusions in the specimen reported turned out to be the mineral magnetite: *this allowed the whole specimen to be attracted to a hand-held magnet.* Remember, though, that serpentine takes many forms and not all will behave in the same way as the piece described.

In *Gems and Gemology* for winter 1995 the GIA report an opaque variegated green-and-white cabochon cemented to the stopper of a snuff-bottle. The immediate impression was of jadeite, but *when tested for RI a 'blink' between two readings of 1.50 and 1.65 was seen. This effect is characteristic of carbonates, which have a high birefringence, and is seen as well-spaced alternating shadow-edges when the stone is rotated on the refractometer.* However, the specimen did not give the expected effervescence with dilute hydrochloric acid, so both forms of calcium carbonate, calcite and aragonite, were eliminated. X-ray diffraction analysis showed that the specimen was a rock consisting mainly of dolomite, some quartz and some other minerals. Dolomite (magnesium–calcium carbonate) reacts to weak hydrochloric acid only when it is powdered.

Carved jadeite can be quite effectively imitated by glass, as reported in the fall 1995 issue of *Gems & Gemology.* A 62.26 ct carved figure, resembling jadeite, showed an RI of 1.56 (jadeite is 1.66), and an absorption spectrum in which absorption extended from 700 to 650 nm and from 480 to 400 nm. *Gas bubbles, a swirly structure and fine crazing-like surface cracks with conchoidal fracture markings showing a vitreous lustre all identified the piece as glass.* Fine detail showed as high relief in this piece: this is not often found in carved glass. Though the carving was moulded, its effectiveness could cause identification problems.

Jadeite assemblages are described in *Gems & Gemology* for fall 1995. A translucent mottled green carving was examined by the GIA, the piece measuring about 33.33 mm long by 22.03 mm wide. A closed setting concealed the depth but a spot reading gave an RI of 1.66, and absorption lines similar to those of chromium could be seen in the red portion of the spectrum. So far the piece could have been jadeite. Under magnification, however, it turned out to be a thin hollow shell of jadeite filled with a transparent colourless material, this containing some gas bubbles. *At one place the filler was exposed on the surface.* Fourier-transform infra-red (FTIR) spectroscopy showed the filler to be a polymeric substance similar if not identical to others previously known to have played this part with respect to jadeite. The thickness of the shell was about 0.05–0.10 mm.

Dyed and carved beads of serpentine were on sale at the 1987 Tucson Gem & Mineral Show. The beads, reported in the summer 1987 issue of *Gems & Gemology,* were partly yellow and partly reddish orange. The beads were made in China and gave vague RI readings of 1.56. The undyed portions fluoresced a moderate dull grey–green under LWUV, the same portion showing a weak dull purplish red under SWUV. The dyed areas gave a variable strong chalky yellow to moderate orange–red under LWUV, the same areas fluorescing a patchy weak orange–red under SWUV. Dye was easily removed with an acetone-soaked cotton swab and one bead, on sectioning, showed that the dye penetrated only a very short distance. The hardness was approximately 4.5.

Plastic imitations of lapis lazuli and malachite are mentioned in the summer 1988 issue of *Gems & Gemology.* They have been used as inlays in watch-face material. Testing is simple since the watches with the natural material feel much heavier than those with the plastic imitation. The unpolished edges between each link of the inlay will appear uneven and grainy where plastic is smooth.

An interesting note in the fall 1986 issue of *Gems & Gemology* cites an imitation of lapis lazuli whose colour could not be removed by the customary acetone-soaked cotton swab, though the beads of the necklace were reported to be staining skin and clothing. *The colour was removed, however, by denatured alcohol, which is found in most scents and colognes.*

In the summer 1986 issue of *Gems & Gemology* the GIA report on a lapis lazuli necklace whose 7 mm violet–blue beads were very deeply coloured. *The beads fluoresced a patchy red under LWUV while under SWUV only some gave the usual chalky green response of natural lapis. The acetone-soaked cotton swab did not remove so much of the colour as expected but paraffin treatment was confirmed by the specimens sweating when tested with the thermal reaction tester. Some of the beads contained a purple dye in cracks, visible under magnification, and the necklace showed a definite brownish red through the Chelsea colour filter – a brighter colour than any natural lapis so far examined by the GIA*. Probably the beads were paraffin-treated, then dyed after the seal created by the paraffin had been removed. The dye was strong enough to cause virtually all of the colour.

In the spring 1991 issue of *Gems & Gemology* an imitation of lapis lazuli is reported by the GIA. The appearance was very like that of the natural material, with an even, dark-violet colour and randomly distributed pyrite grains. *Gemmological testing gave an SG 2.31 (natural lapis is around 2.83), and an RI of 1.55. The stone was inert to LWUV and gave a weak chalky yellow fluorescence under SWUV. The pyrite inclusions stood proud of the surface, showing that they were harder than their host. Some random, shallow whitish areas were also visible. When using light transmitted via a fibre-optic tube, the material passed more light than would be expected from natural lapis; when a Chelsea filter was used with the same illumination the specimen became virtually invisible. Used with reflected light, the filter gave the stone a slightly dark reddish-brown appearance. A weak acrid smell was produced by the thermal reaction tester, with slight melting and whitish discolorations*. These suggest that some form of plastic binder might have been used. In fact the material was barium sulphate with a polymer binding agent, proved by X-ray diffraction analysis.

In the same issue of *Gems & Gemology*, dyed blue calcite marble is shown as a lapis imitation. This item too was a single-strand necklace with uniform beads, believed to be lapis. The RI showed as 1.4–1.6 with a blink suggesting the characteristic birefringence of a carbonate. With a 10 per cent solution of hydrochloric acid (applied in an inconspicuous place), effervescence was seen. Magnesite was eliminated as a possible material since it does not react to a 10 per cent hydrochloric acid solution at room temperature. Colour was removed with an acetone-soaked cotton swab, and looking along the drill-hole a yellow underlying colour was seen. X-ray diffraction analysis showed the material to be a dyed calcite marble.

The use of cosmetics has taken a different turn since the pre-war years when it was more customary to use creams and lotions. In those days, before the natural look was preferred, jewellers and gem testing laboratories were frequently given pearls to clean Today it is sometimes the jewellery that stains the wearer, as the GIA found when a necklace of opaque blue beads with yellow metal spacers was accused of staining its owner. Dyed lapis lazuli was suspected and the absence of fluorescence under UV seemed to reinforce this assumption. The RI, however, was not that of lapis and not all the beads looked the same. As the report, in the summer 1989 issue of *Gems & Gemology* states, some of the beads were found to be dyed calcite while others were dyed jasper ('Swiss lapis'). X-rays were used to test the transparency of the beads, when the calcite was found to be less transparent than the jasper.

Natural sodalite is sometimes mistaken for lapis lazuli and it is not surprising that a synthetic material should be thought worth producing. Samples manufactured in China and reported in the summer 1992 issue of *Gems & Gemology* were found to be heavily included and twinned. The material is colourless as grown, but is irradiated to give the blue colour.

A number of ornamental materials have been coated with acrylic substances to improve their polish: traditionally, wax and paraffin have been used on turquoise and lapis lazuli. In the summer 1992 issue of *Gems & Gemology*, the GIA staff quote an article recommending the use of aerosol sprays for surface improvement. One spray which provided a transparent colourless surface was tested and found to give a glassy coating on fashioned specimens of lapis lazuli and jadeite. Four light coatings were applied, *the surface then showing a concentration of glassy material in irregularities and carved recesses. The coating could be removed quite easily with a razor blade, and also with an acetone-soaked cotton swab.*

A glass imitation of lapis lazuli is noted in the *ICA Laboratory Alert* No. 44 in 1991. Both a bead necklace and a single loose fashioned stone had been examined. The material was reported to be opaque and predominantly medium blue with darker blue portions distributed in a marbled pattern. The RI was found to be 1.62 and there was no response to LWUV though under SWUV a very faint powdery blue could be seen. A uniform distribution of tiny, highly reflective and transparent slightly brown flake-like spots, some with triangular outlines, was seen under magnification. The spots were presumably intended to imitate the brassy yellow pyrite inclusions so often seen in natural lapis.

An imitation of lapis lazuli is reported in the fall 1995 issue of *Gems & Gemology.* The material examined was a pair of scarabs with a blue colour reminiscent of sodalite rather than lapis. *Concentration of the blue colour was found in fractures – a sure sign that the specimens had been dyed.* X-ray diffraction showed that the material was a dyed feldspar.

It is interesting when one substance simulates another which is itself used to imitate something else! In the fall 1993 issue of *Gems & Gemology* this misrepresentation of dyed magnesite as 'howlite lapis' is reported. Howlite is often dyed to imitate turquoise and other coloured gem materials, its fibrous structure making the process easy. The GIA reported large violet-coloured cabochons which were offered as 'howlite lapis'. They were quite convincing as lapis lazuli imitations as they contained white, dye-resistant veining resembling the calcite often seen in natural lapis. *However, the properties turned out to be different from those expected for howlite: there was no expected high birefringence (shown by the characteristic 'blink' on the refractometer). The presence of a dye was established by the use of acetone-dipped cotton swabs.* Further testing by X-ray diffraction analysis identified magnesite. *A mild soap solution removed some of the dye!*

An imitation of lapis-lazuli with an X-ray diffraction pattern matching that for phlogopite (one of the mica group of minerals) suggested to the GIA (*Gems & Gemology,* spring 1993) that it might be a phlogopite ceramic. Further examination of a thin section between crossed polars showed that the specimen was predominantly a strongly birefringent mica-type material with high-order interference colours. Additional observations showed minor singly refractive zones coloured dark blue, these appearing black when the stone was viewed between crossed polars. Testing with a scanning electron microscope with an energy-dispersive spectrometer showed that the specimen was largely composed of crystals with a roughly rectangular outline and lamellar structure, characteristic of mica, the spectrum indicating the presence of magnesium, aluminium, silicon and potassium as major elements, as in phlogopite. Also found to be present were grains of an undifferentiated silicate of calcium and magnesium and of the mineral lazurite.

While turquoise is very often dyed to improve the colour, it is rarer to find a simulated matrix merely painted on to the surface. In the summer 1986 issue of *Gems & Gemology* a porous carved turquoise is described: the specimen had been paraffin-treated and easily reacted to the thermal reaction tester. What made the specimen a little out of the ordinary was the use of a black dye with the paraffin. This 'matrix' was painted on the many flat surfaces of the carving but was not commonly detected in the natural matrix depressions.

While it is usually possible to detect treated and synthetic turquoise with magnification, the spectroscope and the careful use of the thermal reaction tester, backed turquoise, in which the matrix is left with the stone and the piece mounted in a closed setting, can be harder to spot. The GIA (*Gems & Gemology*, fall 1984) state that, if known, the backing should be disclosed as the price is much lower than for unbacked material.

While turquoise is often improved, the normal aim is to stabilize powdery material or to produce a darker blue. A member of the GIA staff on a visit to Egypt was told in Luxor that while tourists preferred blue turquoise the local preference was for green. A turquoise dealer displayed a plastic jar containing a viscous liquid in which several hundred carats of turquoise cabochons were immersed. Apparently mineral oil is boiled for about 1 hour and then allowed to cool back to room temperature. The already fashioned turquoise is placed in the oil and examined daily until the colour change is complete – this takes one or two weeks. The stones are then cleaned with denatured alcohol before sale.

An imitation of turquoise tested by the GIA and reported in the winter 1990 issue of *Gems & Gemology* turned out to be calcite with a plastic binder. *The specimen had an RI of 1.56, a hardness of 2–3 and an SG below 2.57, so that natural turquoise was eliminated.* The quite large piece was a light, slightly greenish blue, with a smooth surface and a moulded appearance on one corner. When a chip was placed in dilute hydrochloric acid, a carbonate effervescence was seen but the chip did not dissolve entirely. After about 15 minutes' immersion the reaction ceased, leaving a rough, whitish surface. This material was easily scraped off. The scrapings when touched with the thermal reaction tester gave off the acrid smell characteristic of plastics: beneath the scraped area the surface reacted once more to the acid. The infra-red spectrum gave a curve indicating calcite – by a coincidence the calcite peaks obscured those given by the plastic.

Turquoise is notably difficult to test and it is all too common to have to turn to sophisticated testing methods for a satisfactory result. A specimen submitted to the GIA in 1994 and reported in the fall issue of *Gems & Gemology* for that year was proved to be turquoise by the standard gemmological tests. *But a colourless transparent material was seen on the surface when the semi-translucent to opaque partially polished blue rough was magnified.* Infra-red spectroscopy showed that this substance was a polymer. Scrapings were then taken from areas of the specimen that appeared to show no polymer: the scrapings were combined with a potassium bromide pellet which, when examined by infra-red spectroscopy, showed that the polymer was also present in the seemingly bare places.

Dyed and impregnated turquoise presented as a slab to the GIA laboratory staff was noted in the summer 1994 issue of *Gems & Gemology.* Normal gemmological testing proved the specimen to be turquoise, but *under the microscope the irregular and very dark-blue periphery showed concentrations of dark-blue material with a subvitreous lustre. The specimen was easily marked by the point of a metal needle, and reacted like a plastic when the thermal reaction tester was brought close.*

One of the problems with YAG is the general absence of inclusions, but some stones, according to a report in the winter 1993 issue of *Gems & Gemology*, are now coming on to the market with at least some useful interior features. A dark-green YAG of 15.45 ct was found to be principally coloured by chromium: inside the stone were elongated gas bubbles inside fine layers coloured blue, and distinct, slightly curved parallel graining. Scattered small crystals were also observed, each crystal surrounded by stress fractures. The stone was either a reject from optical crystal production or a Russian production made by 'horizontal crystallization', a variation of the floating-zone type of growth. If the latter theory is correct, included samples of YAG may become more common.

While green YAG usually does not have quite the colour and appearance of emerald, the unwary could still be deceived. A stone reported in the summer 1992 issue of *Gems &*

Gemology weighed 5.56 ct and gave a negative reading on the refractometer. *The SG was measured at 4.55 (the mean RI of approximately 1.57 and mean SG around 2.68 for emerald easily separate the two specimens). However, in the absence of such gem-testing facilities the 10× lens can always be put into service: in this case the YAG showed natural-appearing inclusions which turned out to be combinations of residual flux and gas bubbles.*

A blue variety of GGG with a colour saturation reminiscent of haüyne is reported by the GIA in the winter 1994 issue of *Gems & Gemology.* The high SG of GGG would identify it should haüyne ever be imitated (unlikely!).

A yellow CZ masquerading as a 10 ct diamond *showed absorption areas close to the 478 and 453 nm bands expected in natural yellow diamond type IA. The line at 415.5 nm could not be detected, fortunately, and the stone fluoresced orange under LWUV. Set with a diamong in a ring, the difference in dispersion between the two stones could be seen*, as reported in the winter 1990 issue of *Gems & Gemology.*

Glass imitations of CZ offered in the United States as 'zirconia' are reported in the winter 1989 issue of *Gems & Gemology.* The pavilions were foiled with a reflective material, giving more brilliance than glass would unaided.

A report in 1989 highlighted the use of CZ crystals as imitations of diamond crystals. Sales were reportedly taking place in Namibia, and prices up to US$4000–5,000 were being asked. Such an imitation would only be saleable in a diamond-producing country, where it is usually illegal for unauthorized persons to own rough diamonds.

A CZ containing inclusions that might have led to misidentification as diamond is reported in the fall 1984 issue of *Gems & Gemology.* The stone, a round 0.90 ct brilliant, *showed irregular swirly growth features somewhat like the graining that may be seen in diamonds. Another CZ seen by the GIA contained spherical inclusions oriented in subparallel lines. The inclusions were seen under magnification to be negative crystals but with voids lined, at least partially, with a white material, perhaps undissolved zirconium oxide. In some of the voids angular growth patterns could be seen.*

When CZ first came on to the market in doped forms to give a range of attractive colours, many thought that the coloured varieties of YAG would be displaced from manufacture. None the less coloured YAG has persisted into the 1990s: in the spring 1992 issue of *Gems & Gemology* a highly saturated 'fluorescent' yellow stone is reported. In a 3.16 ct faceted specimen *the hand spectroscope showed absorption extending from 500 to 400 nm and under both LWUV and SWUV a very strong yellow fluorescence could be seen.* Using EDXRF analysis the colouring agent was found to be cerium. Cobalt-doped specimens gave a medium-dark, slightly violet–blue response, reminiscent of some irradiated topaz. *These stones showed red through the Chelsea colour filter and gave a characteristic cobalt-type absorption spectrum with bands at approximately 640, 595 and 560 nm.* In a greenish-blue stone which looked like some of the blue copper-coloured tourmaline from Paraíba, Brazil, both chromium and thulium were detected. The same elements, with holmium, were found in some green material, but chromium alone was found to be the cause of colour. *These stones will show rare earth fine-line spectra, not seen in any of the species they are imitating.*

A dark yellowish-green CZ, illustrated in the spring 1992 issue of *Gems & Gemology* as a faceted stone of 2.00 ct, *appeared green through the Chelsea colour filter and was inert to LWUV while fluorescing a weak yellowish green under SWUV. With the spectroscope, absorption bands could be seen at 607, 583, 483, 472, 450–443 nm.*

A piece of CZ was used for a carving of Buddha, reports the winter 1993 issue of *Gems & Gemology.* The piece was 30.6 mm high and a transparent brownish yellow colour. Traces of a yellow metal gilt could be seen randomly placed on the figure, said to date from the 16th century, and some foreign material could also be seen in many of the incised areas. *RI testing gave a negative reading, and a bright orange-red fluorescence under both LWUV and SWUV did not identify the material. The absorption spectra of the piece and that of the high lead content glass used in the refractometer were found to be similar and the hardness (tested in an inconspicuous place!) was slightly greater than that of synthetic spinel (8.5). The SG was slightly more than 6.00, the combined tests indicating CZ.* EDXRF analysis showed the presence of zirconium, hafnium and yttrium, thus confirming the tests.

Bicoloured CZ is reported in the summer 1992 issue of *Gems & Gemology*. The GIA saw a crystal weighing 230.10 ct, manufactured in Russia, showing an orange core and a lavender periphery. It was said that the two colours came about through a combination of dopants, cerium oxide (CeO_2) and neodymium oxide (Nd_2O_3). Growth in partially oxidizing conditions also played a part in the coloration. Reduced cerium oxide in the core gave the orange colour, masking the weaker neodymium-caused colour. In the outer section of the crystal the oxidized cerium oxide gave no colour, thus allowing the lavender colour from the neodymium oxide to be seen. *Though not mentioned in the report, the specimen would presumably have shown a characteristic fine-line absorption spectrum from the neodymium.* No natural material was simulated.

A crystal purchased in Bogotá, Colombia, was offered as a blue Paraíba tourmaline. With a greyish-blue colour and a weight of 49.81 ct, the crystal closely resembled the blue indicolite variety of tourmaline found in Brazil. The cross-section showed characteristic three-fold symmetry and roughly parallel striations along its length. Some of the deeper 'striations' contained a reddish-brown earthy staining. The two ends of the crystal had a vitreous lustre, and one showed a large conchoidal fracture. *The RI of 1.52 proved it to be glass.* The crystal is noted in the summer 1992 issue of *Gems & Gemology*.

Imitations of tourmaline from Paraíba have been made from a variety of materials, the aim being to get as close as possible to the very fine green and blue colours of the original stones. One imitation consists of a tourmaline crown cemented to a glass pavilion: this would give tourmaline properties if the crown only were tested. Another specimen was made from two pieces of beryl joined with a bright blue cement. Irradiated topaz that has not been heated after irradiation is also a good substitute. Gemmological testing should account for these materials. Blue apatite cat's-eyes have been offered as Paraíba tourmaline cat's-eyes.

Faceted tourmaline composites with red–white and green sections were reported in Germany in 1990: the stones consisted of portions of differently coloured faceted tourmalines glued together. A composite imitating tourmaline cat's-eye was made by cementing a transparent crown to a fibrous pavilion. *In both cases the cement layer was said to be clearly visible.*

The water-melon tourmaline with a pink core and green 'rind' ought to be a commoner subject for imitation (it is very hard to synthesize tourmaline in gemstone sizes), but imitations do not often appear. In *Gems & Gemology* for winter 1992, however, the GIA report an assembled imitation consisting of an outermost layer made from a veneer of a gem material cut in long narrow slices, two dark blue and two dark green. The slices were striated parallel to the length of the supposed crystal. An RI reading taken on one yellowish-green piece gave 1.63 with a weak birefringence: a very weak absorption line could be seen at 460 nm. The thin slices contained fluid inclusions and internal fractures parallel to their length. The next layer was semi-translucent and seemed to be an assemblage of mineral fragments (perhaps colourless quartz and mica) cemented by a colourless substance containing numerous gas bubbles. The large transparent core seemed on first examination to be pink, but when viewed with a fibre-optic light source it appeared colourless with a dark pink coating. The core was incompletely covered in a few places, and between crossed polars it could be seen to be birefringent. X-ray powder diffraction identified the core as quartz. The piece thus had a rock crystal core, coated with a pink colouring agent and a layer of mineral fragments in cement, covered by strips of tourmaline.

An imitation of a natural tourmaline crystal from Madagascar was found to consist of a tourmaline with concentric zones cemented to a lighter-coloured piece of glass. *The specimen (reported by the German Foundation for Gemstone Research) weighed 61.41 ct and gave an RI of 1.629 and 1.648 for the tourmaline portion and 1.520 for the glass. A distinct boundary plane could be seen under magnification as well as gas bubbles in both the glass and the cement contact zone.*

A cobalt-coloured synthetic spinel examined by the GIA and reported in the summer 1991 issue of *Gems & Gemology* showed *misty stringers and veils as inclusions, effects not seen before in a synthetic spinel. The stone, weighing 2.51 ct, was an attractive blue, and gave an RI of 1.728 with red fluorescence under LWUV and a chalky yellow glow under SWUV. These features matched with cobalt-coloured synthetic spinel, as natural cobalt-blue spinels are normally inert under UV.*

An interesting composite of a type not previously reported is shown in the fall 1995 issue of *Gems & Gemology*. This was a 'gem construct', consisting of sections of rhodolite (garnet), colourless topaz and iolite. *The specimen, resembling a prismatic crystal, was thus encircled by bands of different colours*. While the quartz was synthetic, no one could possibly class the whole piece as either natural or synthetic, and no natural stone could have been in mind!

While the mineral actinolite plays a number of roles in gem materials, it is not usually thought of as an ornamental species in itself. In the winter 1993 issue of *Gems & Gemology* a 12.25 ct dark-green tablet is reported, the specimen showing planes of coarse fibres suggesting natural actinolite. *The RI was 1.60 and the SG 2.72. The specimen was in fact a natural glass, a rare example of one natural material imitating another.* Natural actinolite could be used ornamentally but probably would not have much greater a value than natural glass: this specimen showed devitrification which was responsible for the bladed effect.

The mineral zincite (ZnO) is rarely if ever seen as a gemstone in jewellery though it is sought by collectors for its fine deep-red colour which occurs when a trace of manganese is present. Zincite has been synthesized for research and industrial purposes, and since crystals produced are of facetable size and because doping can produce attractive colours, cut stones have been sold to collectors who may believe that they are natural. Most of the synthetic products are pale green or pale yellow in colour.

In *Gems and Gemology* for spring 1995 the story is taken in an interesting direction. Faceted stones showing fine orange and light green colours had been on sale, their origin stated to be the result of long-term deposition in the air vents of a chimney of an industrial kiln used in the manufacture of paint, in which zincite is an important constituent. There have been reports in the mineralogical literature of zincite being formed as the result of mine fires but crystals have been small. The Polish material was sold as faceted stones and as crystals, and showed enough birefringence to allow doubling of back facets to be seen under magnification: the SG was in the range 5.68–5.70 with moderate to very weak yellow–orange fluorescence under both types of UV: interestingly, the strength of the response was inversely proportional to the depth of the body colour of the stone. *While faceted zincite could possibly be mistaken for other rarely cut materials, it is unlikely to be mistaken for diamond despite its high lustre, since it is much denser and the birefringence is detectable without difficulty.*

Synthetic zincite grown hydrothermally or by the vapour growth method is reported as a transparent yellow faceted stone of 4.05 ct in the winter 1985 issue of *Gems & Gemology.* The polariscope gave a uniaxial interference figure while no RI could be read on the standard refractometer. The SG was measured at 5.70, and the hardness determined at 4–4.5. No absorption spectrum was visible. Various colours have been induced by appropriate dopants.

An iridescent imitation of hematite is described in the summer 1995 issue of *Gems & Gemology.* Marketed as 'iridescent specularite', the black material had a near-metallic lustre, with an iridescent surface coloration (interference colours) appearing on oxidation. The oxidation layer could be removed during the polishing process. Found as a hematite-rich rock (slate or shale) from the area of Prescott, Arizona, the material was probably left to weather once mined or is placed in a bucket of water with an iron nail for company. *There should be no difficulty identifying specimens since there would be no reason for false description.*

A simulant of haematite (the natural material is heavy with an SG of 5.20) has been made from silicon produced from a refined melt by the pulling method. *Gems & Gemology* for spring 1991 describes the material, trade-name 'Hemalite': *it has less than half the density of hematite and has an SG of 2.33.* The colour is medium grey with a metallic lustre, and the hardness is 7 compared to 5–6 for hematite.

Rhodochrosite, though comparatively soft, can be very beautifully coloured, and such material cannot escape imitation. In the summer 1990 issue of *Gems & Gemology* the GIA report a massive calcite dyed to resemble the pink–orange–red of rhodochrosite. *Dye was found concentrated in drill holes and in surface-reaching fractures: it was removable with an acetone-soaked cotton swab.* An acid test would not have distinguished between the two species.

A man-made crystal which has been given the imaginative name Oolongolite was produced in the late 1980s by Dominique Robert of Lausanne, Switzerland. Reported in the winter 1989 issue of *Gems & Gemology,* the material has been marketed in a variety of cuts. Colours include blues and greens of different strengths: colourless specimens are also reported. *Gemmological tests showed that the material had a hardness of 7.5–8, an SG of 6.7–7.0 and an RI 1.93–2.00.*

Synthetic periclase (magnesium oxide, MgO) has been used as a faceting material as well as in luminescence studies. In the *Australian Gemmologist*, Vol. 19 (1995), the cathodoluminescence of this white or pink material is offered as a possible means of identification. Pink stones were found to give a crimson response, and white specimens a bright blue response. Some specimens responded with an orange–red cathodoluminescence. You have to think about cathodoluminescence testing at an early stage!

GIA staff have reported an yttrium aluminium perovskite (YAP) doped to give several different colours. The natural material is a calcium titanium oxide: *since the dopants I have seen give a rare earth absorption spectrum, gemmologists should have little difficulty with this material.*

In the winter 1994 issue of *Gems & Gemology* the GIA report on a Russian-made material with the trade name 'Minkovite'. Two stones received from an American dealer weighed 1.19 and 4.50 ct and were a round brilliant and an emerald-cut, respectively. Both were a dark, saturated slightly violet–blue colour and reminded staff of cobalt-doped synthetic spinel. The RI was found to be 1.785, 1.788 and 1.810 for the biaxial negative stones, with a birefringence of 0.025. The specimens were strongly pleochroic with the colours light blue, slightly greenish-blue and violet–blue. A visible–UV absorption spectrum consistent with neodymium doping was seen and EDXRF analysis showed that the stones contained yttrium, silicon and neodymium. The material was determined as monoclinic synthetic yttrium silicate (Y_2SiO_5). The stones displayed curved colour banding, and the larger specimen contained irregular wisps of dark blue colour concentration with one white acicular and several small angular inclusions.

Phenakite is not usually grown for ornamental use, but *Gems and Gemology* for fall 1994 reports a collection of 10 transparent blue–green crystals and crystal fragments, produced by flux growth in Russia. A crystal which could have produced a faceted stone of around 2 ct *gave properties normal for phenakite and showed a very distinct dichroism of dark bluish–green and light bluish–green. Flux inclusions proved the stone to be synthetic.*

The material Tavalite, reported in the summer 1986 issue of *Gems & Gemology*, has been found to be CZ with a thin optical coating to give a different appearance in transmitted and reflected light. Examination showed a characteristically dusty surface and a build-up of coating at facet junctions. Gentle cleaning should not disturb the coating.

The fairly rare mineral amblygonite, normally a pale lemon yellow in colour, has been irradiated to give a pale green colour. This is not likely to affect the price of the stones, which are collectors' items rather than jewellery stones. It is possible that 'amblygonite' is really montebrasite, as the two closely related species are often confused and have a similar appearance.

The popularity of the beautiful blue zoisite variety tanzanite made it a natural object for imitation, and it is surprising that a reasonable look-alike took so long to appear on the market. In the fall 1995 issue of *Gems & Gemology* the GIA report a Verneuil-grown synthetic sapphire with a tanzanite colour. The stones weighed from 0.97 to 4.07 ct, and the trade names Cortanite and Coranite have been used; stones labelled 'tanzanite' and sold in India have been found to be synthetic corundum, whose properties are easily investigated. While the SG and RI were normal for corundum, the stones had been found to contain iron, titanium (the two elements needed to obtain a blue) and a very small amount of chromium (0.003 wt%). The reaction to both types of UV radiation was not strong and a better test is to *look for the curved colour banding; the stones also showed notably dark crowns with lighter pavilions*. A faint red transmission was observed and the 'tanzanite' blue was probably obtained by placing the optic axis at different angles to the table, never at right angles. It was noticed that the smaller the angle of the optic axis to the girdle, the more violet the stone. Despite the smallness of the chromium content, it is thought that it still plays an important part in coloration.

An imitation of tanzanite is described in the summer 1996 issue of *Gems & Gemology*: the faceted transparent stone was found to contain yttrium, aluminium and europium with an orange–red colour seen through the Chelsea filter and faint curved striae visible in one direction. Under SWUV the stone gave a moderately chalky strong reddish–orange fluorescence, which would be sufficient to distinguish it from tanzanite. Absorption lines were observed at 589, 530, 480, 472 and 468 nm, and the stone had an SG of 4.62 with a negative reading on the refractometer.

Topaz has occasionally been used as part of a composite stone, and a recent report in the winter 1995 issue of *Gems & Gemology* highlights a 1.48 ct faceted oval imitation of the 'Paraíba' tourmaline, which in its natural state is coloured by copper to give a superb bright blue. The natural stones sell for very high prices and it is not surprising that imitations have turned up.

In this composite, colourless topaz, cheap and easy to obtain, forms both the crown and pavilion. The blue colour is obtained from the greenish-blue cement holding the stone together. *Any fine blue tourmaline should be examined from the side*, when the deception should be obvious. If the setting prevents this and you are suspicious, the refractometer will show a *birefringence of only 0.01*, much too low for tourmaline, which has a mean birefringence of 0.018, enough to show doubling of inclusions and of back facet edges viewed through the table. *Gems & Gemology* states that no synthetic tourmaline has yet (1996) been made. Tourmaline is a name given to a group of individual mineral species in which a number of chemical elements play a part in formation and coloration. Tourmaline, like other silicates, cannot be grown by the cheap flame-fusion method easily, if at all, and while it might be possible for gem-quality and gem-sized crystals to be grown by another method, the price of natural stones is probably insufficiently high for serious research work to be profitable.

At the 1992 Tucson Gem & Mineral Show an irradiated green topaz was on sale, the product being advertised as 'Ocean Green Topaz' and apparently emanating from Sri Lanka.The colour ranged from light to medium tones and from yellowish and brownish green to more saturated green to slightly bluish green.

While malachite is not a rare mineral, colour-enhanced specimens have been reported. A cabochon-cut stone examined by the GIA and reported in the Fall 1995 issue of *Gems & Gemology* showed a fibrous texture along the length of the cabochon. The specimen showed an RI of 1.58 on the flat base with no sign of birefringence. Readings of different parts of the top showed either vague edges just below 1.60 or a very weak birefringence blink (two shadow edges, widely separated and very easy to spot when the specimen is rotated on the refractometer) from 1.66 to a negative reading. The SG was hydrostatically determined at 3.76, and under LWUV the top of the stone fluoresced a very weak and mottled chalky blue, the lower portion giving a similar but stronger effect. *Under magnification a transparent colourless material was seen to be forming a layer on the base and partially filling a large cavity. This material was most prominent in the areas giving the strongest response to LWUV. Infra-red spectroscopy showed results similar to those obtained for the epoxy resin Opticon.* The GIA report that Opticon has in fact been used to treat malachite from the Morenci copper mine in Arizona. The rough material is first heated in a toaster oven for a few hours at 150–250°F (65–120°C): it is then removed and coated with Opticon and returned for reheating, perhaps overnight. The material is then slabbed and processed again in the same way, this sequence being repeated during the preforming and pre-polishing stages. Multiple low-temperature heating of this kind apparently rules out the need for a chemical catalyst in the form of a polymerizing agent.

Malachite has not only been synthesized but is often imitated – odd, since the material is neither difficult to obtain nor expensive. In the winter 1994 issue of *Gems & Gemology* the GIA report on two pieces of jewellery. One was a necklace with a number of bezel-set stones measuring 11 × 9 mm, showing prominent banding ranging from medium to pale green. The RI was determined at 1.55 and the material effervesced slightly with a weak solution of hydrochloric acid. There was a conchoidal fracture, and a weak green fluorescence under LWUV with a fainter green under SWUV. The SG could not be determined as the stones were set. Small bubbles and the swirly appearance of one of the bands, with the slight amount of effervescence, suggested a manufactured material. With the thermal reaction tester the material turned to a chalky white and flowed easily away from the test point. The verdict was an imitation of malachite.

The other piece was a polished trapezoid measuring 18.0 × 3.8 mm set in a bracelet with obsidian, sodalite, imitation sugilite and imitation turquoise. The stone showed swirly background colour banding and dark-green spots on the lighter bands. Up to this point the stone could also have proved to be a malachite imitation, but a birefringence blink and strong effervescence with weak hydrochloric acid indicated that the stone was probably malachite. It was inert to both types of UV. The thermal reaction tester produced a small brown spot rather than the chalky white with flow seen on the imitation.

The mineral berlinite has been synthesized for various research and industrial applications, but it is rare to find it used ornamentally. In the summer 1989 issue of *Gems & Gemology* this aluminium phosphate is mentioned with respect to its 'bull's-eye' interference figure, which is identical to that shown by quartz, with which it is isostructural. Berlinite is not likely to be found in contexts where confusion with quartz might be expected, but it is useful to know that the figure is not peculiar to quartz after all.

The ornamental minerals charoite and manganoan sugilite have been imitated by a composite consisting of massive beryl with intergrown quartz. The material is heated and then dyed purple; the heating increases porosity, allowing the dye to enter more easily, to a depth of at least 0.5 mm. A turquoise-blue and coral orange–red material has also been produced by a Swiss manufacturer, as reported in the summer 1992 issue of *Gems & Gemology.* High prices were at one time asked for the best purple translucent specimens of manganoan sugilite.

Gems & Gemology for summer 1995 reports a metallic ornamental material which has been given the trade name Platigem. Few metals appear in the gemstone role, and this one is an intermetallic compound of platinum, aluminium and copper. The GIA examined five faceted pieces ranging in weight from 1.09 to 12.63 ct. The colour was light yellow to brownish pink with an even distribution and metallic lustre. No RI reading could be taken on the refractometer, the material was inert to both types of UV, was not magnetic and showed no absorption spectrum. The SG was between 8.66 and 8.77, and the hardness of one specimen was about 5.5. Specimens appeared to be isotropic. *Gemmologists should have no particular trouble in identifying the material from these notes and in any case there are no real imitations of Platigem!* The material is made from the brassy yellow intermetallic compound $PtAl_2$ mixed with 5–25 wt% copper (this gives the colour). There are some structural similarities with fluorite.

A cameo made from ceramic alumina had a hardness of 9 and was coloured white on blue. Qualitative chemical analysis showed the presence of alumina and cobalt, the latter being responsible for the blue colour. The advantage of such a material lies in its durability compared to genuine cameos made from shell.

An imitation cat's-eye with the trade name 'Fiber Eye' is reported in the summer 1991 issue of *Gems & Gemology.* Available in either white or brown, *stones show the back facets projected on to the crown as the fibres are oriented perpendicularly to the table*: the effect is reminiscent of the natural mineral ulexite ('television stone'). The specimen in question was faceted especially for testing, however.

A drilled glass greenish-grey chatoyant bead was examined by the Bahrain Gem and Pearl Testing Laboratory and reported in the April 1995 issue of the *Journal of Gemmology.* The bead gave a vague RI reading of about 1.62. Viewed in the direction of the drill-hole the bead showed a very pronounced hexagonal pattern, and *it was possible to read print through the bead, as in the natural mineral ulexite.*

Chapter 14

Glass, ceramics, plastics, composites and experimental materials

Glass

I have often begun a gemmology class by saying that most gemstones are glass, and if you take the total carat weight of all natural, synthetic and imitation stones this is very likely to be true! Anyone who has looked at thousands of stones will be able to develop an instinctive feel for glass, based on a number of factors: an unexpectedly light weight for the species the glass is imitating; a 'glassiness' arising from an absence of mineral inclusions; when a suite of jewellery is examined, stones closely matching in colour are usually glass (though a Verneuil synthetic is also possible); unusual colours are also more likely to be glass than anything else. Two other general points: imitation pearls are most usually glass; the term 'paste', with an interesting history, means glass to the gemmologist and jeweller of today. All paste is glass.

The chemical and physical nature of glass is complex, and simple explanations do not really cover the facts. For our purposes we need only to know that glass has no regular internal atomic structure as all crystals do and that therefore it cannot have those directional properties seen in the majority of (but not all) crystals. These include directional hardness, pleochroism (difference in colour shown in different directions), birefringence and the electrical properties of pyroelectricity and piezoelectricity. Glass is a poor conductor of heat compared to any crystalline substance and will feel warm when lightly touched with the tongue. Such a test is not recommended for general use, however, since a specimen may have previously been immersed in liquids, or even tested on the refractometer, without being adequately cleaned.

Glass shares with minerals of the cubic system the absence of birefringence since light travels at the same velocity in glass whatever direction the light ray takes. For the gemmologist relying on the 10× lens, doubling of the back facets cannot be seen and since singly refractive gemstones are well in the minority, glass is well to the front of possible identities for an isotropic (singly refractive) unknown. Once inside the glass with the lens, an unmistakably swirly structure can be seen and, usually, large, well-rounded gas bubbles randomly distributed.

Most glass contains silica and the glass you see in windows is usually sodium–calcium glass. Various additives can be used for different purposes, including lead to give a high dispersion and boron or aluminium oxide to increase heat resistance and resistance to some chemicals. The glass used in the refractometer is a lead glass with a refractive index (RI) of 1.962, and lead glass is also used for glass ornaments confusingly named 'crystal'. Thallium is also used to produce a glass with high dispersion. A wide variety of colouring agents is used, ranging from gold to copper or selenium producing red, cobalt oxide producing blue and uranium compounds giving a rather startling yellow–green colour which is easily recognizable. Nickel can give a bright apple green while iron can give a variety of colours; whether the glass is manufactured in oxidizing or reducing conditions also has an effect on colour. Transparent glass is naturally in great demand, while the use of opacifiers produces a translucent or opaque material.

The cheaper glass gemstones are moulded rather than faceted, and such specimens show unmistakably rounded facet edges. Sometimes a compromise is found in stones where the table facet only is polished after the stone has been removed from the mould: tin oxide is commonly used as the polishing agent.

If the lens does not give a conclusive result on a glass, remember that very few facetable natural minerals are singly refractive and give an RI between 1.50 and 1.70 (both the natural possibles, pollucite and rhodizite, are colourless). The hardness of glass is usually between 5 and 6.5. Glass is brittle and shows multiple conchoidal fractures on girdle and facet edges: this type of fracture is also common in natural materials but occurs much less frequently.

Between crossed polars on the polariscope glass shows very marked anomalous double refraction, showing a striped effect which is highly characteristic. Since glass is made in such widely differing compositions it is not surprising that some specimens may show effects which are hard to distinguish from those shown by natural materials: some glass may show a dark cross-like shape which might cause confusion, but rotation of the specimen will not cause the cross to alter in position and in no case will there be regular changes from light to dark, nor complete darkness during a complete rotation.

A glass with the beryl composition and emerald coloured from doping with chromium has been manufactured by melting and cooling the appropriate ingredients. The properties are slightly lower than those of beryl, and various names have been used for it, beryl glass being the commonest. The RI is close to 1.52 and the specific gravity (SG) is usually about 2.42. The addition of cobalt gives a blue colour (the aquamarine-like result has been called mass aqua) and doping with didymium gives pink – and a recognizable rare-earth absorption spectrum.

An excellent imitation of cat's-eye chrysoberyl has been made by manufacturing thin glass rods, which can form masses of hollow parallel tubes. When a cabochon is formed from this material, light reflected from the tubes gives a reasonable eye, and the body colour of the cabochon can be chosen at will. Even more convincing is 'Cathay stone', made by making thin glass optical fibres in fused mosaics. The fibres, made from several distinct glasses and stacked in cubic or hexagonal arrays, can give a very sharp eye as there can be as many as 150 000 per square centimetre. Neither of these products will give a chrysoberyl absorption spectrum (a strong band in the deep-blue region near 440 nm) nor the appropriate RI from a flat base.

Different fluorides and phosphates can be added to glass to imitate moonstone or pearl. Again, the product differs sufficiently from the natural to allow easy distinction to be made.

In the 1970s a partially crystallized glass with the name Victoria-stone, Meta-jade or Kinga-stone was manufactured in Japan. Made in a number of different colours, the materials show different amounts of devitrification (the precipitation out of some constituents of the glass), and in some examples fibrous inclusions have been added to give chatoyancy. Iimori of Tokyo, the manufacturers, have also made a clear or translucent material in a variety of colours, said to imitate jade.

Slocum stone has been mentioned in the section on opal. It is a glass in which tinsel-like laminated material has been included, with a spacing of about 0.3 μm between the laminations allowing diffraction of light to take place at the inclusions, which in some cases resemble the colour patches in opal, which also arise from diffraction. The usual glass features of swirls and gas bubbles can be seen under magnification.

Aventurine glass or goldstone is made with cuprous oxide, and is reduced during annealing to give tiny flakes of metallic copper: the aim is to imitate the sunstone variety of feldspar. A blue opaque to translucent glass containing copper flakes may imitate lapis lazuli.

A glass with a fine emerald colour with an RI of 1.635 and an SG 3.75 was found to be radioactive. It showed absorption from 700 to 600 nm and from 440 to 400 nm with two lines centred at 470 and 460 nm. The radioactivity was found to be twelve times the background level.

Ceramics

Ceramics are defined by Nassau (1994) as finely ground inorganic powders which may be heated, fired or sintered and sometimes compressed to give a polycrystalline solid. The particles may be held together by a binder with a low melting point, and the surface is often glazed. The composition may resemble that of lapis lazuli (see Chapter 13), turquoise or jade. The Gilson lapis material, described under lapis lazuli, is a good example of a ceramic gemstone imitation. Ceramic alumina has been offered as a cameo material and one specimen, with a white on blue composition, had a hardness of above 8 for the white portion. The blue portion was found to have been coloured by cobalt.

A ceramic material with the trade name Yttralox and the composition Y_2O_3 has been used to imitate diamond. It is completely transparent and has a hardness of 6.5, an RI of 1.92 and a dispersion of 0.039.

Plastics

Plastics are commonly used to imitate a variety of gemstones but more especially the organic ones. Many of them are examined more closely in Chapter 12 on organic materials. Most plastics are soft, between 1.5–3 on Mohs' scale, and have an SG of 1.05–1.55. The RI is usually about 1.5–1.6, but many specimens may be adversely affected by the refractometer contact liquid.

An early plastic used for ornament was the familiar celluloid, a mixture of the lower nitrates of cellulose and camphor heated under pressure to 110°C. Early specimens were highly inflammable: the use of acetic acid removes this risk. Cellulose acetate ('safety celluloid') has a hardness of 2 and an SG of 1.29. 'Old' celluloid has an SG of 1.35, with fillers rising to 1.80. The hardness is 2 and the RI 1.49–1.52. Safety celluloid burns or chars with a vinegary smell while the majority of plastics when heated give off a characteristic acrid smell which irritates the nose. Specimens are not of course held in a match or candle flame but an instrument known as the thermal reaction tester (formerly 'hotpoint') is brought close and the effect on the eyes or nose observed. Another type of plastic made in a variety of quite attractive colours is casein, formed from the protein part of milk and hardened by the addition of formaldehyde. Casein has an SG in the range 1.32–1.39 (usually 1.33) and an RI of 1.55. If a drop of concentrated nitric acid is put on the surface a yellow spot will develop: this somewhat destructive test is perhaps best left to the textbooks, and the thermal reaction tester used instead to produce a burnt milk smell.

Bakelite is a phenolic resin and is inclined to become yellow with age. Less tough than some of the other plastics, it has a wide colour range and an SG of 1.25–1.30. The RI is 1.61–1.66. If chips of Bakelite are immersed in distilled water in a test-tube and the water is then boiled, a small amount of 2,6-dibromoquinonechlorimide can then be added. On cooling, a drop of very dilute alkaline solution is added, and if a blue colour forms, the presence of phenol is indicated, proving the specimen to be Bakelite. This destructive test is not likely to be needed by the jeweller who should prefer to become familiar with the materials imitated by Bakelite and the other plastics. Amino plastic is a modification of Bakelite, and being transparent can be dyed to give a range of colours. The hardness is close to 2, and the SG close to 1.50. The RI is 1.55–1.62.

Probably better known than some of the above is Perspex, an acrylic resin with a low SG of 1.18 and RI of 1.50. It is used for cheap beads (which will not lie properly when made up into a necklace, compared to pearl). Polystyrene resins are moulded to resemble faceted stones and have an SG of 1.05 and an RI of 1.59. Di-iodomethane and other organic liquids will easily dissolve these resins.

Glass, ceramic materials and plastics usually declare their nature to the experienced gemmologist and do not often need to be tested to find out what they actually are. While the number of genuine gem materials imitated is large, there is often no serious resemblance and plastics in particular are so light that suspicion should instantly be aroused.

Composite stones

The name composite is used to describe gemstones which are made from two or more materials that may be foreign to or sometimes the same as the stone imitated: thus a diamond doublet is made from two pieces of diamond while a garnet-topped doublet is usually glass with a thin slice of almandine fused to the table (Figure 14.1). The secret of spotting composites is to 'think composite' all the time when examining a set of unknown specimens: some are very convincing (we all know the opal doublet) and, when unset, can easily pass even the experienced eye. The older textbooks spoke of 'true doublets' and 'semi-genuine doublets' when the parts were from the same species and when one part was from the species imitated, but these terms are falling into disuse, if they ever had much currency.

While composites are made to deceive there is 'deceit and deceit'! Manufacturers of doublets (they do not really exist!) would say that a diamond doublet (these are very rare, and I have seen only three in many years) can use pieces of diamond to make a larger stone than would otherwise be possible and that larger coloured stones (I have never seen a coloured diamond doublet) can be made from composites. Another argument is that a garnet-topped doublet can give you better and longer wear than glass alone (but why not buy a genuine stone?). It was often asserted in the older textbooks that the fusion of a slice of almandine with a piece of glass was to resist testing with a steel file, but this cannot have been a regular test since stones with pronounced cleavage (diamond in particular) would easily be damaged in this way.

The opal doublet, described in Chapter 11, used to avoid wasting very thin opal of fine quality, is simply opal on opal. Ruby on ruby and emerald on emerald have been reported but are certainly rare. A jadeite doublet was said to consist of a thin layer of green jadeite on a thicker layer of white jadeite, the whole cut as a tablet.

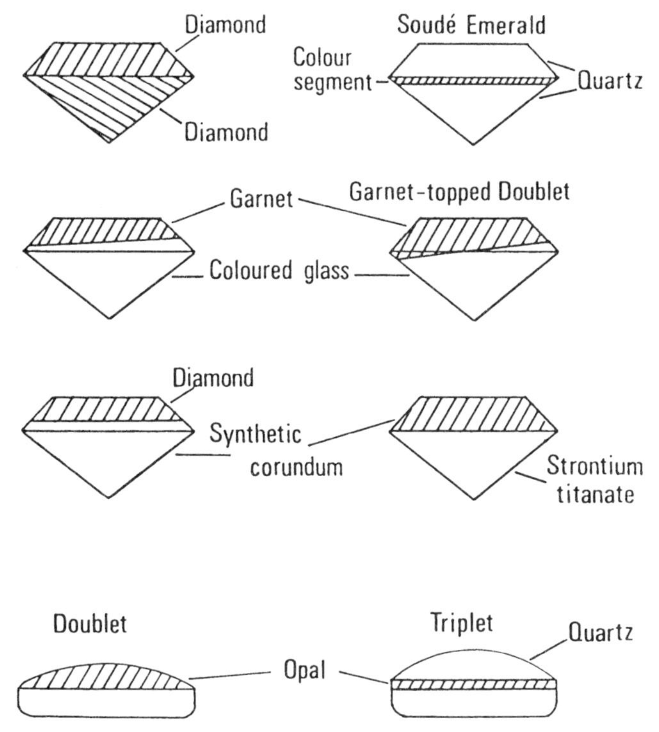

Figure 14.1 A selection of composite gemstones. After P.G. Read)

The diamond doublet can be spotted by the reflection of the table a little way down into the stone, and the effect is even better seen when a pencil point or other object is placed on the table. Immersion can also reveal the two sections, which may even come apart if the test is prolonged. Some Indian jewellery contains what have been called lasque diamonds: these are thin parallel-sided plates with the top often faceted. The shadow of the table edge may be seen reflected from the lower surface: pieces set with such stones should be carefully examined, but Indian jewellery should always be tested with this possibility in mind. This is not a deceit but a use of material which happens to be found in this form. A diamond interior showing rainbow-like colours should be examined to see if they arise from thin films of air trapped between two portions of a doublet or if they denote internal cracks or cleavages.

While almost any combination of doublet is possible, awkward specimens include pale emerald on green glass, included aquamarine on green glass and a highly transparent chrysoberyl cat's-eye on a darker base of unknown composition. Specimens of natural sapphire on synthetic sapphire or ruby can give trouble: the presence of natural inclusions in the crown and a characteristic absorption spectrum from the base could be deceptive, even though Verneuil blue sapphires tend not to show the otherwise characteristic absorption band at 450 nm. If care is taken with the microscope the curved growth lines should be seen or if an ultra-violet lamp is to hand, short-wave radiation will show the characteristic greenish glow associated with synthetic blue sapphire (a better test than looking for the elusive 450 nm absorption band). If the base is ruby, long-wave radiation will show its bright red glow, which will contrast with an inert top of other red material. A one-off (it is hoped) consisted of a quartz top glued with a green cement to a green-dyed base, the whole thing offered as green jadeite: this would lack the jadeite absorption band at 437 nm.

Rock crystal is easily available, is clear, hard and has no easy cleavage. This makes it an ideal top for a doublet, and leaves the manufacturer with a wide choice of base, which will most often be glass. When such a composite is in a closed setting it could escape detection, and only if the gemmologist or jeweller detects a colour abnormality will the deception be found. While textbooks usually quote SG figures, I have yet to find anyone who regularly removes stones from their setting for such a cumbersome test to be carried out: it is much easier, if the stone is coloured, for the spectroscope to be used or, failing a result, the refractometer. As I have said elsewhere, the majority of jewellers want to know if their stone is a ruby or emerald (and so on): if it is not one of the usual jewellery stones they will not be concerned over its true identity. Some interesting examples cited in *Webster's Gems* (Robert Webster, revised Peter G. Read, 5th ed., 1994, Butterworth-Heinemann) were sapphire blue, purple and yellow. The blue and purple specimens showed a cobalt absorption spectrum: this was a good thing since the stone resembled amethyst and would, with a rock crystal top, have given the correct RI for amethyst! Yet again, a mistake was saved by a spectroscope.

The yellow stone, resembling a yellow synthetic spinel (these stones are a rather vivid yellowish green colour and fluoresce a strong yellow under LWUV) fluoresced yellow from the base only, as you would expect, and the fluorescent glow when examined with a spectroscope showed bands ascribed to uranium. This effect is known as a fluorescence spectrum, and uranium glass, green or yellow, does turn up quite often. In the case of the yellow doublet, the cement layer showed as a bright line surrounding the girdle.

Just to make things a little more fun, this type of doublet can be reversed with a quartz base and a glass top. Another example described in *Webster's Gems* (1994) was a white topaz with the tip of the pavilion made from natural blue sapphire. In many natural Sri Lankan blue sapphires a spot of blue in the base suffices to turn the whole specimen blue (the cause can be seen from the side) but this did not seem to work with the topaz. A very rare type of doublet may be filled with a coloured liquid (reported by Bauer in *Edelsteinkunde* (1896), and thereafter): the stone may be rock crystal or glass, and is hollowed out with a high polish given to the walls of the cavity. (This must have been slow and expensive and no wonder examples are rare.) The coloured liquid was placed in the cavity and the stone completed with a base of the same material as the crown. I am sure few if any examples are still around in jewellery – what did the liquid consist of and did it degenerate? Testing is obvious (or should be).

Colour is not always a guide to composites. The well-known garnet-topped doublets are most commonly green or blue, but pink and red ones are not too rare. A specimen reported in *Webster's Gems* (1994) was red and consisted of two pieces of colourless glass joined by a coloured cement. The RI was 1.52 and the SG about 2.48. The red cement was said to show an almandine (i.e. iron) absorption spectrum!

A garnet-topped doublet should be one of the easiest imitations to diagnose since the lens or microscope may show characteristic mineral inclusions in the garnet layer: these will most commonly be needle-shaped crystals of rutile. Before the lens is called into play, look for unexpected flashes of red in a specimen which is not red (even colourless garnet-topped doublets have been reported, although I have not seen one). Moving a stone – any stone – in the beam from a fibre-optic light source is often enough, and if the stone is looked at against a white background, and especially if it is placed table facet down, a red ring will be seen encircling it.

If a garnet-topped garnet remains unsuspected and gets as far as the refractometer or spectroscope, anomalous readings (or lack of them) should give the game away. As almandine has too high an RI for the jeweller's refractometer, the gemmologist will immediately see that the green stone cannot be peridot nor the blue stone a sapphire, both species giving a clear reading: care is still needed since the garnet slice may not be almandine with the highest possible value for the species and you may just get a reading. Should the spectroscope be the instrument of choice (always recommended for coloured specimens) the strong iron absorption band at 505 nm will be seen: it is not easily confused with the bands which the imitated species would show.

Sometimes part of a specimen placed under UV will show an unexpected greenish-white response while the remainder will show none: a garnet-topped doublet is a possibility since quite a lot of glasses may respond to SWUV while iron-rich almandine will not.

Although we have mentioned soudé stones under emerald (see Chapter 9), as composites they deserve a place here too. Two pieces of rock crystal joined by a green gelatinous substance were probably the earliest emerald-imitating composites, and can be spotted by normal testing, but if you rely on the Chelsea filter beware, since these stones show red, though not strongly: over time some of the green material may have yellowed. Other colours have been reported: an alexandrite imitation gave a cobalt absorption spectrum. Watch for the quartz RI and, should you be using UV, for a suspicious glow from the cement layer.

Soudé stones are still turning up, with better-quality coloured material used for cementing the two portions. Some of the later specimens remain green under the Chelsea filter but the quartz readings are, of course, unvarying. It is possible that the joining layer may be a coloured lead glass, which would account for a slightly higher SG: this also seems to be used in a composite in which the two main portions are made from synthetic spinel. Manufactured in France from 1951, the specimens give the expected RI (1.728) from colourless synthetic spinel, and glow a sky blue under SWUV. Immersion, as always, reveals the deception, with liquids of a higher RI than water being especially useful. Under the Chelsea filter the green soudé spinels remain green while blue zircon and aquamarine imitations show green: when a solid synthetic spinel is made to imitate these two species, it will usually show orange through the filter. Soudé spinels imitating amethyst and sapphire show orange, however. In general, the Chelsea and other filters should not be relied upon without the back-up of other instruments: none the less, they give an early clue to these particular composites.

Other examples of composites will be found in the chapters dealing with particular species, and in the reports from recent literature. Remember especially the use of strontium titanate as the base of small-sized diamond imitations: these are some of the most deceptive composites I know, and their smallness often ensures that testing is not considered. Remember, too, the dangerous jadeite triplet, the various opal composites and the odd imitations of star stones. Finally, in *Webster's Gems* (1994) is reported a mosaic composite in which two colourless materials (often synthetic spinel) sandwich a three-colour mosaic in a transparent filter. I do not know what the product looks like as this is not stated, but it illustrates the lengths to which experimenters will go!

Reports of interesting and unusual examples from the literature

Items in this section have been chosen to illustrate points made in the chapter and to bring one-off items to your notice.

In the spring 1992 issue of *Gems & Gemology* a chatoyant glass is described with a more natural-looking 'eye' than that seen in previous specimens. *An 11.00 ct cabochon showed the chatoyant band intersected at right angles by a series of evenly spaced dark lines.* Using the microscope and looking from the side perpendicular to the 'eye' revealed a honeycomb structure with the individual cells displaying hexagonal outlines. Individual fibres were thicker than in some imitation cat's-eyes, and their edges were not transparent, leading to a diminution in chatoyancy precision. *An RI taken on the dome of the cabochon showed two readings, of 1.48 and 1.62.*

Glass eggs are common and not normally expected to show any particularly unusual features. This is not always the case, however, as the Gemological Institute of America (GIA) found on examining a green glass egg weighing 198 g. Gas bubbles and swirls established the egg as glass but the piece showed a very strong greenish-yellow fluorescence. A test for radioactivity (made because the egg resembled some uranium glass known to be radioactive) showed a reading ten times above the normal background radioactivity.

Glass manages to turn up in a number of roles. In the spring 1987 issue of *Gems & Gemology* a light green emerald-cut stone is reported. *It was found to have a single RI of 1.529 and to show a singly refractive reaction between crossed polars but showing a strong anomalous double refraction with a straight parallel pattern corresponding to features seen in the same position when the stone was examined under the microscope. There was a very weak dull-yellow fluorescence under LWUV and a very weak chalky greenish-yellow response under SWUV. No absorption was detected with the hand spectroscope.* The specimen was identified as glass, but the resemblance to a light-green beryl or unheated aquamarine was very close. The stone was large enough to be hefted to give some idea of the SG, which appeared to be quite close to that of beryl. Heavy liquid testing in fact gave an SG of approximately 2.50.

While most glass used to imitate gemstones has an RI in the range 1.50–1.70 and a common SG range of 2.30–4.00, the GIA report in the winter 1993 issue of *Gems & Gemology* on a material with the trade name 'Junelite'. This has been produced in a range of colours, and the publicity material accompanying the stones (seen at a gem show) quoted an SG of 4.59 and an RI of 2.0. *On testing, the SG was found to be 4.44 and the RI above 1.81 (the limit of the refractometer contact liquid used at that time – it is now 1.79). The material displayed anomalous birefringence between crossed polars*, and its amorphous nature was confirmed by X-ray powder diffraction.

I myself have been shown glass masquerading as rough emerald. A hexagonal crystal with unusually smooth faces was brought to me with the statement that the owner had invested his life savings (derived from farmland in Central Africa) in these stones. The crystal had flakes of a supposed mica adhering to it but it was not the usual biotite mica, which is dark brown to black. When examined with a strong light, large gas bubbles could be seen inside the 'crystal', which was thus proved to be glass. Fortunately the owner did not seem to be too upset by my verdict!

A green glass reportedly fashioned from ash ejected by the eruption of the volcano Mount Saint Helena has been on sale for some time. The claim is that the transparent glass has resulted from fusion of the very fine ash, but this claim is refuted by Nassau in the summer 1986 issue of *Gems & Gemology*: such fusion produces a black glass with a different composition from the green material. The glass was still on the market by the publication of the fall 1992 issue, and indeed in 1996! Some at least of the glass has been sold as 'Emerald obsidianite'. The publicity material accompanying the glass states that the green colour is due to traces of chromium, iron and copper.

A jadeite doublet is described in the fall 1986 issue of *Gems & Gemology*. The specimen appeared a fine green but consisted of a very thin green layer on the top and another, thicker white layer beneath: thicknesses were 0.1 and 2.2–2.3 mm, respectively. *The layers were joined by a slightly yellowish cement containing numerous gas bubbles.* The dark-green upper layer was mottled and contained many nearly colourless veins: chloromelanite was suggested. The white lower layer had a distinct crystalline structure. *RI readings were 1.64–1.74 on the green layer, which showed characteristic chromium absorption lines with the hand spectroscope.* Both layers were identified as jadeite by X-ray diffraction analysis, but the variation in RI of the upper layer remains unexplained.

Emeralds which were in fact rock crystal with a green backing were found by the GIA in a segmented reversible necklace. The heart-shaped green stones gave an RI reading of 1.54, too low for emerald, and were seen to contain two-phase hexagonal inclusions. The stones were in a closed setting and it was not possible to determine the nature of the backing. The necklace was quite elaborate, and proves the point that even expensive-looking jewellery may not be all that it seems.

A German firm was reported to be marketing beryl triplets in a very good emerald colour. In this product care had evidently been taken to use beryl pieces with characteristic inclusions, and some pieces had been carved as cameos. The same firm was offering beryl triplets with a saturated, slightly greenish-blue colour as an imitation of the blue Paraíba tourmaline.

Star doublets are mentioned in gemmology textbooks but are not very common. The GIA report on a transparent red cabochon, bezel-set in a man's ring, and showing a six-rayed star. *The RI and other features showed that the crown was a Verneuil synthetic ruby, and under magnification round and oval gas bubbles could be seen in the cement layer. The base of the cabochon, which was a reddish-purple colour, showed strong hexagonal growth zoning, partially healed fractures and other signs suggesting it to be natural corundum of the low-quality type sometimes known as 'mud ruby'.* The report, in the fall 1993 issue of *Gems & Gemology*, goes on to say that a positive identification of the base portion was not possible since the setting prevented it. The rutile crystals ('silk') in the base did, however, provide the star effect in the crown.

Faceted tourmaline composites with red–white and green sections were reported in Germany in 1990: the stones consisted of portions of differently coloured faceted tourmalines glued together. A composite imitating tourmaline cat's-eye was made by cementing a transparent crown to a fibrous pavilion. *In both cases the cement layer was said to be clearly visible.*

A triplet imitating emerald and made from synthetic spinel and glass is reported in the winter 1986 issue of *Gems & Gemology*. The GIA report that this type of triplet was first made in 1951 by Jos Roland of Sannois, France, where the stone was called *soudé sur spinelle*. Stones were made in various colours by sintering coloured glasses to the colourless spinel crown and pavilion. The stone examined by the GIA weighed 11.66 ct and was emerald-cut. Between crossed polars the triplet had a cross-hatched appearance and curled black bands, effects characteristic of synthetic spinel. Interestingly, the RI taken from the table showed a reading of 1.724 (strong) and a weaker one at 1.682.

Between the two readings some other shaded areas could be seen. From the 0.55 mm thick green glass layer an RI of approximately 1.682 was obtained. The hardness of this layer was about 4, a figure shared with many other highly refractive glasses. No absorption bands could be seen with the hand spectroscope. *Under the microscope small flattened, rounded and irregularly shaped bubbles could be seen in the separation plane: these showed best when fibre-optic illumination ws used. Looking perpendicularly to the girdle, the thick glass could be seen to show rounded edges and very prominent swirls. Immersion in di-iodemethane showed the composite nature of the specimen very clearly. Under LWUV the crown showed a strong chalky yellowish-white fluorescence when viewed nearly perpendicular to the girdle with the table closest to the UV source; the glass layer was inert and the pavilion gave a strong clear yellow fluorescence with no trace of chalkiness. When the culet was placed close to the source an opposite effect could be seen, with colours reversed.* The GIA believed that this effect was caused by the glass diminishing the amount of radiation reaching those parts of the stone that were not directly facing the source. Under SWUV the stone was virtually inert, and with X-rays a very weak chalky green fluorescence was seen. There was no phosphorescence.

It is not easy to imagine gemmologists being taken in by this ingenious composite but as always, with an unknown behaving oddly, you have to 'think composite'. Specimens of the same type have been reported in yellow, purple and orange–red colours.

Experimental materials

Continuing the gemstone manufacturing story, we have to look at a number of materials which have been grown for research and industrial purposes and which, if appropriately fashioned, may make beautiful gemstones, often with magnificent colours or with high dispersion. I freely acknowledge that such products are often small or soft or cleave too easily and, yes, I know that they are *rare* and very difficult and expensive to obtain. None the less they may turn up as gemstones, and we should also remember that there are collectors who specialize in these often exceptional products.

For reasons other than their ornamental qualities, several crystalline substances grown for research are diamond-like, not always because they are colourless but because they may be brown or yellow, with high dispersion and single refraction. Most such materials can only be tested by gemmological instruments in such a way as to show that they are not diamond and if one is encountered in jewellery conditions this would suffice. However, the jeweller just *might* come across a specimen in the vicinity of Murray Hill, New Jersey, Malvern, Oxford or Cambridge, England! Some crystal growers have been known to have their products fashioned and mounted!

Germanium shares some physical properties with silicon and has substituted for it in some artificial substances. There are two chemical forms of *bismuth germanate*, and both can be grown as large transparent crystals which may be colourless or, as seems more common, a rich brown, orange or yellow colour with high dispersion which inevitably suggests diamond. Crystals are not hard (4.5 on Mohs' scale would be a typical value) so jewellery use is ruled out: however, if ever available they would be keenly sought by collectors. The SG is high, over 7, which would make a fashioned specimen noticeably heavy. Perhaps on appearance alone, a faceted specimen would most closely resemble sphalerite, zinc sulphide, often collected.

Germanates have been doped with rare earths, which would give the characteristic absorption spectrum. A *bismuth silicate* has been grown in colourless or orange to brown forms. Bismuth does not feature in the natural gem mineral world.

The beryllium oxide *bromellite* comes into every gem reference book, so here it is again: do not cut it, do not lick it (if in dust form) – it has a high toxicity, and is rarely cut for this reason, despite its hardness of 9 on Mohs' scale. Reported faceted stones are colourless.

A possible diamond simulant is *hafnia*, the hafnium analogue of zirconia but heavier and more expensive. I have seen one colourless crystal grown by the skull-melting method. Somewhat more common is *yttrium orthoaluminate* ($YAlO_3$), also possible as a diamond simulant but frequently discoloured by impurities: it has a hardness of 8 +, an SG of 5.35 and an RI of 1.938. Specimens would show up with the reflectivity meter but only as non-diamond. Also involving yttrium is the oxide Y_2O_3 with a hardness of 7.5–8, and SG of 4.84 and an RI of 1.92. The dispersion is high at 0.039. The name *Yttralox* was coined for this material, but it was never commercially significant.

Recently back on the market (if it was ever on it before) is *silicon carbide*, highly dispersive but also highly birefringent. Its green colour could easily be mistaken for the colour of green diamond, if you were prepared to overlook the birefringence (0.043). The hardness is over 9, and the SG is 3.20. The RI is 2.648 and 2.691 – this is very high for a mineral. Under LWUV stones have been reported to show a mustard-yellow fluorescence.

Doping with rare earths and other elements is always indicative of artificial origin and we have already seen many examples. Yet another is *zincite* (ZnO), which has been grown in colourless and orange, yellow or pale green forms. This material is currently of particular interest because fine, large faceted orange and yellow stones have been on the market in limited numbers. They are said to have formed as a by-product of the manufacture of other materials and to have been recovered from a flue in a factory. Zincite is known to have occurred as a product of mine fires (in zinc mines, of course) and is also grown hydrothermally!

Natural colourless *scheelite*, calcium tungstate ($CaWO_4$), has been mistaken (rather than used) for stones of higher value, including diamond. Scheelite is manufactured for a variety of industrial purposes, and large clear crystals have been faceted. These, like their natural counterparts, fluoresce a strong sky-blue colour under SWUV (but remember some of the synthetic gem diamonds): fortunately for the diamond trade, scheelite is strongly birefringent. Dopants are sometimes added.

Just for interest, *zircon* (zirconium silicate, not zirconia) has been grown, and one or two beautiful purple crystals, showing prism and pyramid forms and with strong pleochroism, are recorded. I mention this because there are gem crystal collectors who could very easily be taken in (you believe what you want to believe!): natural coloured zircon is not usually this colour, nor does it show strong pleochroism. The purple colour is obtained by doping with vanadium. Crystals doped with terbium have also been seen: they are yellow and quite small.

Also doped with vanadium is the beryllium silicate *phenakite*: examples I have seen are slender light-blue crystals with smaller crystals growing from the prism faces. Specimens are most attractive, and would certainly appeal to the collector were they available. Bell Laboratories grew them, as they grew many other beautiful and interesting materials. The magnesium oxide *periclase* has been faceted and marked under the trade name Lavernite: specimens are colourless with an RI of 1.73 and SG of 3.5–3.6. Some stones show a whitish glow under UV.

While natural *fluorite* (CaF_2) is quite often cut as a gemstone, despite its relative softness and easy cleavage, a synthetic counterpart has occasionally been faceted. One example in the literature is colourless and shows a green fluorescence under LWUV. Another is a brilliant green with an exceptionally long phosphorescence under X-rays: the dopant was said to be indium. I have seen a red crystal of fluorite in which uranium was said to be the dopant. From the literature, another red uranium-doped fluorite showed a rare earth absorption spectrum with a particularly sharp line at 365 nm. Yet another red synthetic fluorite contained a variety of cavities, straight growth planes and crystallites, combining to give a natural appearance.

For the benefit of crystal collectors, the following turn up from time to time: I have seen all of them over the years. Testing does not usually turn out to be necessary since they are usually encountered in research laboratory or university conditions to which few gemmologists have access. Some crystals may always sneak out, however!

The mineral group of *apatites* is quite complicated and has a number of individual mineral members: some are silicates rather than the more familiar phosphates. Both classes have been synthesized and dopants have been freely used. The most likely variety to surface is the neodymium-doped lilac to purple type with a change from one colour to the other in daylight and incandescent light, respectively. Two analogues of strontium titanate – which we have already met – are the calcium titanate *perovskite*, doped forms of which are currently on the market, and *barium titanate*, which has not so far been commercially produced. Both materials have higher SG values than diamond and both are appreciably softer.

Greenockite, the transparent orange-coloured cadmium sulphide found only as small crystals in nature, has been grown to quite large sizes using the vapour phase growth method. The colour of faceted stones quite resembles the colour of some diamonds, and the high RI of over 2.5 (diamond is 2.417) helps to give a high degree of lustre. The absorption spectrum shows virtually complete absorption below 525 nm, an effect which would not be seen in diamond.

Analogues of strontium titanate, already mentioned as a reasonable though soft diamond simulant, are *lanthanum niobate* and *neodymium niobate*: though not produced as diamond simulants (the latter is coloured and soft) they may turn up as interesting crystals. The material known to crystal growers as 'banana' (*barium sodium niobate*) is colourless, as is *potassium niobate*. The best known niobate is the highly dispersive but birefringent *lithium niobate 'Linobate'*, now rare.

As well as germanates (there is a pink *erbium germanate* with an erbium rare earth absorption spectrum) and niobates, an *yttrium vanadate* has been grown in a form sufficiently clear to be used ornamentally.

Among the rarer synthetic garnets, mentioned earlier, are amethyst-coloured *neodymium gallium garnet* (these are oxides rather than silicates) and yellow *samarium gallium garnet*. Other than fluorite, some analogous materials include *manganese fluoride* and *rubidium manganese fluoride*, both pink, and chromium-doped yellow *lithium fluoride*. *Magnesium fluoride* crystals in a pale-green colour have been produced by the Massachusetts Institute of Technology. Apart from the last, the spectroscope will at least show that the other fluorides are not natural minerals.

Molybdates do not normally feature in gemstones (except as a possible flux ingredient). The mineral *wulfenite*, lead molybdate, is usually too small to be used ornamentally though it has a magnificent orange colour: however, a colourless synthetic counterpart has been grown in large sizes and shows very high dispersion, though it is soft. Analogous is a *calcium molybdate*, the natural mineral *powellite*: this has been doped with rare earths to give a pink colour, probably among others. Despite a low hardness of 3.5, collectors may try to obtain the material because it shows a variety of fluorescent effects.

The silver arsenic sulphide *proustite* is a superb dark red and would make a magnificent gemstone were it not for its softness and proneness to developing a surface alteration on prolonged exposure to light. It would not be appropriate to facet natural crystals, but an artificial form has been grown by the Czochralski method. This is a really beautiful synthetic material which would be keenly sought by collectors.

I have sometimes been asked why there are no gem-sized synthetic topaz and tourmalines. The reason is that both are quite complex multicomponent substances and hard to grow: we should also remember that the name tourmaline is given to a mineral group, in the same way as the garnets and feldspars, and that the members of the group have a complicated chemistry. Additionally there has to be some research, industrial or commercial reason for investigation and finally growth to be undertaken. Neither topaz nor tourmaline is geologically rare.

Among non-transparent materials, *azurite* and *malachite* have been synthesized, but the natural minerals are plentiful. In the case of *haematite* the same is true, but none the less a haematite imitation is on the market (there are several in fact): while the natural mineral is well-known for its red streak, the imitations do not show such an effect. *Titanium dioxide*, *compressed lead sulphide powder*, and *'hemetine'* (a mixture of stainless steel with chromium and nickel sulphides) have all been used to imitate haematite. At least some of the imitations are magnetic, a property not shown by haematite. Another dark and magnetic material sometimes made artificially is the iron oxide *magnetite* and a rarer example is black *magnetoplumbite*, which has a brown streak.

If you have a chance to read a catalogue of a crystal-growing facility (such catalogues are as rare as some of the crystals!) or to consult the *Journal of Crystal Growth* or the *Materials Research Bulletin*, you will see that the number of substances that have been grown at one time or another is amazingly large. Only a very small number of the possible compositions have been grown for ornamental use and, as a matter of interest, the number of possible inorganic substances, natural or artificial, is limited by the properties and behaviour of the chemical elements.

I have not detailed possible tests for most of the materials we have just surveyed since they are either obvious (usually the spectroscope) or the specimens themselves do not lend themselves to the rather robust methods of gemmological testing. None the less, the selection is important since we can learn how and sometimes why crystalline substances are grown – but the study is quite a long way from simple gem testing and readers do not have to worry about encountering such materials on a seriously large scale.

Details of them and of other growth methods excluded from this text can be found in the journals just cited – but it is fair to say that there is quite a jump from simple gemmology to the understanding of some of the chemistry and thermodynamics involved! In most everyday circumstances of gem testing such studies are not necessary, and nothing will be lost by not knowing about them. If, on the other hand, you have, for some reason, to find out about them, such study is degree level rather than that of the gemmological diplomas, so if the latter is your highest point, there is still plenty to learn!

It is a great thing when this adventure is discovered.

Index